T0324746

Petrochemical Economics

Technology Selection in a
Carbon Constrained World

CATALYTIC SCIENCE SERIES

Series Editor: Graham J. Hutchings *(Cardiff University)*

Published

CATALYTIC SCIENCE SERIES — VOL. 8

Series Editor: Graham J. Hutchings

Petrochemical Economics

Technology Selection in a Carbon Constrained World

Duncan Seddon

Duncan Seddon & Associates Pty Ltd, Australia

ICP

Imperial College Press

Published by

Imperial College Press
57 Shelton Street
Covent Garden
London WC2H 9HE

Distributed by

World Scientific Publishing Co. Pte. Ltd.
5 Toh Tuck Link, Singapore 596224
USA office: 27 Warren Street, Suite 401-402, Hackensack, NJ 07601
UK office: 57 Shelton Street, Covent Garden, London WC2H 9HE

British Library Cataloguing-in-Publication Data
A catalogue record for this book is available from the British Library.

Catalytic Science Series — Vol. 8
PETROCHEMICAL ECONOMICS
Technology Selection in a Carbon Constrained World

Copyright © 2010 by Imperial College Press

ISBN-13 978-1-84816-534-2
ISBN-10 1-84816-534-X

Printed in Singapore.

To Joan & George

PREFACE

The objective of this book is to give industry professionals, engineering, research scientists and financiers an overview of the technologies and economics for the production of olefins in the petrochemical industries. The book gives an overview of the options and costs for producing olefins using different technologies and from different feedstocks at a time when the cost of carbon dioxide emissions are set to be included in the processing cost.

The book is based on a series of workshops and specialist seminars given by the author dating from 1996 in Singapore and Kuala Lumpur. The workshops focussed on production economics, improving plant profitability, feedstock supply and cost. The book is an updated and expanded version of the author's workshop notes.

The book critically compares the alternatives so that the most attractive options for petrochemical production can be identified for specific locations and conditions. For this purpose, technology capital and operating costs have been compared on the same basis (US Gulf location to a late 2007 cost base). From this, the production costs are estimated for various feedstock prices and compared to the traded prices of the products where appropriate.

The contents are widely embracing as possible for viable technologies in 2008. From time to time new technologies are identified or more information on emerging technologies become available. For brevity, the book does not cover technology still in the research and development stage. In particular, the extensive volume of material on the direct conversion of gas (methane) into ethylene has been omitted.

The text is roughly divided into two parts: the first six chapters discuss steam cracking technology and the approaches to olefin production from hydrocarbons and the later chapters concentrate on the production economics.

Units

A technical and economic appraisal of petrochemicals spans several large subject areas: petroleum and oil industry economics, petrochemical refining and applied chemistry, chemical engineering and process economics. Unfortunately these distinct fields carry their own units. The petroleum industry generally uses American units based on standards defined at 60° Fahrenheit and are generally the units used in the US chemicals industry. Most chemists and academic engineers use S.I. units which are the most widely used units used in the European chemical industry. However, much petroleum engineering and refining technology uses a bastardised version - often mixing American and S.I. units in the same function.

The book generally employs S.I. units which the author considers to be the most widely acceptable. American units are used where appropriate. All costs are in US dollars of late 2007. Note the possibility of confusing US one thousand (M) with the S.I. Mega (M). In this work Mt is million metric tonnes and MM$ is million US dollars.

Economic Analysis

The economic analysis follows the methodology described by the International Energy Agency for evaluating alternative feed stocks for the production of petrochemicals. Full details of the method are given in the Appendix. In summary the issue is that petrochemical technologies are highly capital intensive and some means has to be found for comparing alternatives. The method used for technology comparisons is to develop a fixed variable equation for a hypothetical green-field plant producing olefins from a given feedstock. The fixed costs of the plant are derived from published estimates of the costs of the capital items and operating costs. The operating costs can often be approximated as fixed percentages (typically 10%) of the total installed capital cost. The return on working capital is also included in the fixed cost term. No account is taken of tax, depreciation or allowances.

This leaves the feedstock as the only variable in the fixed-variable equation:

$$P = x.F + C$$

where P is the product production cost, F is the feedstock price, x is the variable constant and C is the constant representing the fixed costs of capital and non feedstock operating costs.

The largest component of the constant C is the return on investment of the fixed capital. In order to make comparisons easier, a standardised methodology is adopted which is detailed in the Appendix. Typically process plants are assumed to take a similar (3 year) construction period to operate at full output over the project lifetime with zero residual value. The capital payback is then over this operating life.

Once the fixed variable relationships are derived the equation can be used to estimate the production cost for any given feedstock price. By comparing the estimated production cost with traded prices for the product, the viability of a particular project can be determined. By considering alternative technologies at similar feedstock prices, alternative approaches can be critically compared.

Approximations to the Economic Analysis

Reference is made to figure the Figure A below.[1] This figure illustrates the cost error for any given project as the project proceeds to completion. The error plus or minus is the error from the final cost which is only known after the project is complete.

The first stage in the project is the concept study. This involves minimal expenditure in terms of the total project cost. The error in the cost estimate ranges from about +/-25% to +/-40% of the final project cost. The primary aim of this work is to improve the approach to the concept study to achieve an error in the lower end of this range. For many occasions (e.g. for very remote or unusual locations) this may not be feasible and the errors may be as much as 100% or more.

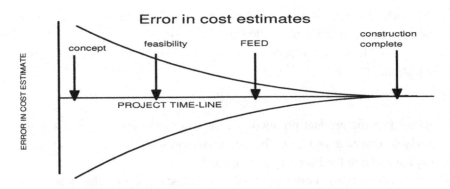

Figure A: Error in an estimate as a function of project life

The second phase is the feasibility study. This stage may require the expenditure of 1 to 2% of the total project cost. Thus for a $1,000 million project about $20 million will be required. This will define the location, feedstock and product market and the technology to be used. It will also typically encompass outline regulatory approval and assessment of environmental impacts. The error in the estimate is typically not less than about +/-10%. Financiers (bankers and corporate boards) like the error to be +/-5%. This level of estimate can usually only be achieved by a FEED study.

Front-End Engineering and Design (FEED) focuses on obtaining accurate cost estimates for the major items of process equipment and further definition of the most sensitive parts of the overall project cost. It is denoted Front-End because this is performed prior to final corporate board and financial approval for the overall project. FEED costs typically 5 to 10% of the overall project cost – thus for a $1,000 million project expenditure of about $100 million will be required. Only after the feasibility and FEED will the cost error be in the vicinity of +/-5%. In many instances, especially for new technology or for a remote location or offshore, the error will still be in the region of +/-10% or more.

The full engineering design, procurement and construction will also account for some error, hopefully <5%. Only when the project is completed and started and running to the design specifications is the final cost of the project known.

As the adage goes, "the accuracy of the cost estimate is proportional to the time and money expended." Unfortunately there is a common tendency to try to shorten or circumvent the costing process, which often leads to project failure.

Data Sources

The world petrochemical industry is surveyed annually in the *Oil & Gas Journal* as the "Ethylene Report." This is a useful source of country production, individual steam crackers (including ownership) and the feedstock used. Since 2006 US olefins and the US natural gas liquid supply and prices are each reviewed twice per year by Lippe.[2] Weissermel and Arpe[3] have provided an excellent description of many technologies and approaches to chemical synthesis in the chemical process industry.

In this book, wherever possible literature references are given which should be followed for further information. The *Oil & Gas Journal* articles are a useful source and these often give further references to conference proceedings and articles published in the academic literature. As well as technical articles, *Hydrocarbon Processing* produces reviews of technology on a regular basis. Nowadays, these are issued on a CD ROM and provide more details of different technologies from the various process licensors.

In the descriptions of the various technologies, several assumptions and omissions to the process flow sheets have been made in order to help understanding of the principal issues and to improve the clarity of the descriptive. If a particular technology or approach is of interest to the reader then the process licensor should be approached for the latest updates and information.

For many chemicals reporting agencies such as ICIS-LORS and Platts produce daily price and volume bulletins for subscribers. ICIS-LORS data is reported for a wide range of chemicals regularly in ICIS Chemical Business (formerly European Chemical News). Other groups such as Chemical Market Associates regularly report on global trends and prices which are often reviewed in the *Oil & Gas Journal*.[4]

Purvin & Gertz Inc. produce regular reports concerning the LPG trade. Some of these reports are reviewed in the *Oil & Gas Journal*.[5]

Oil statistics, natural gas and propane prices are readily available from the US Energy Information Administration website (www.eia.gov) which as well collating a vast amount of current and historical data offers useful links to other sites.

---

[1] See also P.B. McIntire, *Oil & Gas Journal*, Aug 13, 2001, p. 30

[2] For example D. Lippe, *Oil & Gas Journal*, Jul. 7, 2008, p. 64; *idem.*, Nov. 3, 2008, p. 54

[3] K. Weissermel and H.-J. Arpe, "Industrial Organic Chemistry", VCH Publishers, New York, 2nd edition 1993

[4] For example M. Eramo, *Oil & Gas Journal*, Dec. 5, 2005, p. 52 and see also *ibid.*, Aug. 25, 2008, p. 48

[5] For example W. Hart, R. Gist, K. Otto. D. Rogers, *Oil & Gas Journal*, Jun 23, 2008, p. 58

CONTENTS

CHAPTER 1

WORLD ETHYLENE PRODUCTION BY STEAM CRACKING

The world ethylene production capacity is approximately 120 million tonnes (2008)[1]. The regional break-up is shown in Figure 1.1.

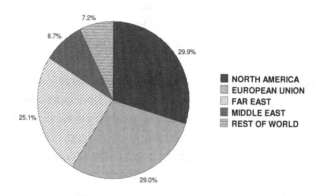

Figure 1.1: World ethylene capacity (120 million tonnes 2008)

In 2008 the ethylene production capacity was still dominated by the developed economies of North America, the European Union and the Far East. The Far East is dominated by Japan and Korea but with significant contributions from the countries of South East Asia. Emerging and rapidly growing regions of olefin production are China and the Middle East.

The following is a selected review of the world's major cracking operations producing olefins and petrochemicals.

1

North America

USA

The North American production is dominated by the very large cracking operations in the USA reflecting the United States position as the single largest petrochemicals market. It has a large number of fully integrated plants producing a comprehensive range of petrochemicals. In 2008, the US's capacity was almost 29 million tonnes per year (t/y) which is 80% of North America's operations and 24% of the world's total.

Production in the US is on a par with the Far East which has recently overtaken the USA in nameplate capacity. The USA is also slightly larger than the expanded European Union which has major integrated petrochemical operations in Belgium, The Netherlands, Germany and the UK.

Although the USA is geographically large, the petrochemical operations are concentrated in Texas and Louisiana. This gives them easy access to the large oil and gas production facilities in Texas and Oklahoma and the growing production of oil and gas from the Gulf of Mexico. This geographical concentration also facilitates the interchange by pipeline of chemical intermediates (ethylene, etc.) and the development of large open markets for such interchange.

In the past, natural gas liquids – ethane, propane and butane – were the favoured feedstock for ethylene production. Propylene was extracted from the off-gas of some of the world's largest oil refineries in the same region. In recent times, naphtha crackers and flexible fuel crackers have been built (the favoured approach in the Far East and Europe). However, as the following Figure 1.2 illustrates, natural gas liquids (ethane, propane and butane) account for the major portion of the ethylene feedstock.

The product slate from cracking natural gas liquids is dominated by ethylene. Propylene in the US is made from refinery off-gases (REF GAS) and there is a small contribution to ethylene from this source as feed to ethylene cracking operations. Naphtha makes up the balance and

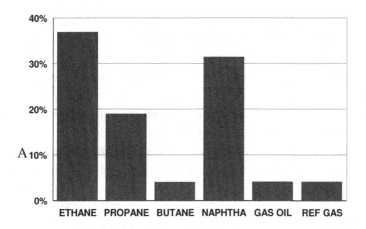

Figure 1.2: US ethylene feedstock (2008)

again much of this is sourced from the natural gasoline fraction of natural gas liquids (condensate).

The approximately 40 cracking operations are owned by various corporate entities. Some have several plants across the US. As well as US majors (Chevron-Phillips, Exxon-Mobil, Dow Chemical, Equistar), several foreign organisations operate crackers in order to have better access to the US market. Notables amongst these are BASF-Fina (EU), Formosa Petrochemical (Taiwan) and Sasol (South Africa).

Of the more than 40 US cracking operations, most are world scale with an average capacity of over 700,000t/y. The US has some of the largest plants in the world with several in excess of 1 million tonnes and one with over 2 million tonnes capacity. Table 1.1 gives a list of the ethylene cracking operations, the operators, their location and nameplate capacity in 2008.

As well large integrated plants for producing olefins and resins, the US chemical complexes can source large volumes of aromatics and other chemicals from the juxtaposed refinery operations. Because these are some of the largest refineries in the world, speciality products can often be simply extracted at a minimum cost. A good example is the production of linear paraffins (for the production alpha-olefins, which are used to produce biodegradable detergents). In most parts of the world this is a costly exercise, but these important intermediates can be extracted in the volumes required from the jet-fuel stream of the large Texas refineries (e.g. Exxon-Mobil's Baytown refinery has a capacity 523,000bbl/d). The large integrated and open market with many competitors, easy access to low cost engineering contractors and equipment, large operations which maximises the economy of scale and access to low cost feedstock has developed the Gulf region of the USA as a centre for low cost production of petrochemicals.

The only issue of concern is the reliance on the ready supply of natural gas liquids and the price of the gas used in their production.

Table 1.1: US Ethylene Plants and Capacity 2008 (tonne/year)

COMPANY	LOCATION	t/y
BASF FINA Petrochemicals	Port Arthur, TX	830000
Chevron Phillips Chemical	Cedar Bayou, TX	794000
Chevron Phillips Chemical	Port Arthur, TX	794000
Chevron Phillips Chemical	Sweeny, TX	923000
Chevron Phillips Chemical	Sweeny, TX	673000
Chevron Phillips Chemical	Sweeny, TX	272000
Dow Chemical	Freeport, TX	630000
Dow Chemical	Freeport, TX	1010000
Dow Chemical	Plaquemine, LA	520000
Dow Chemical	Plaquemine, LA	740000
Dow Chemical	Taft, LA	590000
Dow Chemical	Taft, LA	410000
Du Pont	Orange, TX	680000
Eastman Chemical	Longview, TX	781000
Equistar Chemicals LP	Channelview, TX	875000
Equistar Chemicals LP	Channelview, TX	875000
Equistar Chemicals LP	Chocolate Bayou, TX	544000
Equistar Chemicals LP	Clinton, Iowa	476000
Equistar Chemicals LP	Corpus Christi, TX	771000
Equistar Chemicals LP	Laporte, TX	789000
Equistar Chemicals LP	Morris, ILL	550000
ExxonMobil	Baton Rouge, LA	975000
ExxonMobil	Baytown, TX	2197000
ExxonMobil	Beaumont, TX	816000
ExxonMobil	Houston, TX	102000
Formosa Plastics Corp. USA	Point Comfort, TX	725000

Table 1.1 (continued)

Formosa Plastics Corp. USA	Point Comfort, TX	816000
Huntsman Corp.	Odessa, TX	360000
Huntsman Corp.	Port Arthur, TX	635000
Huntsman Corp.	Port Neches, TX	180000
Ineos Olefins and Polymers	Chocolate Bayou, TX	1752000
Javelina	Corpus Christi, TX	151000
Sasol North America	Lake Charles	453515
Shell Chemicals Ltd.	Deer Park, TX	1426000
Shell Chemicals Ltd..	Norco, LA	900000
Shell Chemicals Ltd..	Norco, LA	656000
Sun Co. Inc.	Marcus Hook, PA	225000
Westlake Petrochemicals	Calvert City, KY	195000
Westlake Petrochemicals	Sulphur, LA	567000
Westlake Petrochemicals	Sulphur, LA	522000
Williams Energy	Geismar, LA	612245

However, since 2000 gas prices have spiralled resulting in increased feedstock costs. In recent times, on an energy basis, the cost of gas in the US has often exceeded the cost of crude oil. This has led to the erosion of operating margins for the large number of plants using gas liquids as feedstock. It has also facilitated exports to the US from low production cost operations elsewhere, such as the Middle East.

Canada

Canada with a nameplate production capacity of 5.53 million tonnes of ethylene is a major player in world petrochemicals. Most of the product is devoted to the US market, particularly the northern states which are remote from the integrated operation of the US Gulf. Most of the operations are large gas based operation based in Alberta: Table 1.2.

Mexico

Mexico has a nameplate ethylene capacity of 1.384 million tonnes. This is produced by three operation owned by Petroeleos Mexicanos. All operations use ethane feedstock.

Table 1.2: Canadian Petrochemical Operations

COMPANY	LOCATION	t/y	FEEDSTOCK
Dow Chemical	Ft. Sask. ALTA	1285000	Ethane
Imperial Oil Products	Sarina, ONT	300000	Ethane LPG
Nova Chemicals	Corunna, ONT	839002	LPG naphtha
Nova Chemicals	Joffre, ALTA (E1)	725624	Ethane
Nova Chemicals	Joffre, ALTA (E2)	816327	Ethane
Nova Chemicals	Joffre, ALTA (E3)	1269841	Ethane
Petromont	Varennes, QUE	295000	LPG naphtha

European Union and Russia

The countries of the European Union (EU) have cracking operations with an annual nameplate capacity of about 26.4 million tonnes of ethylene (2008). The breakdown across the E.U. is shown in Figure 1.3.

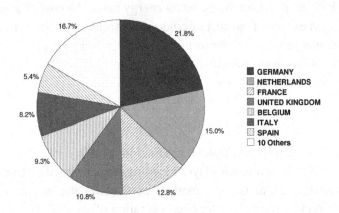

Figure 1.3: EU ethylene capacity (26.4 million tonnes 2008)

The pie chart (Figure 1.3) shows that the major operations are in Germany France, the Benelux countries and the UK. Like the US cracking operations, which are near refineries, oil and gas producing facilities of Texas, Louisiana and the US Gulf, many of the EU petrochemical centres are juxtaposed to refinery operations, North Sea oil & gas producing centres and major ports. This gives them feedstock integration with refinery and natural gas production.

In contrast to the US, most of the feed used in the production of petrochemicals is naphtha with a minor portion coming from natural gas liquids (ethane, propane etc.). There is a minor contribution (just below 10%) from gas oil, much of this being waxy residual fuel oil. The breakdown of feedstock is shown in Figure 1.4.

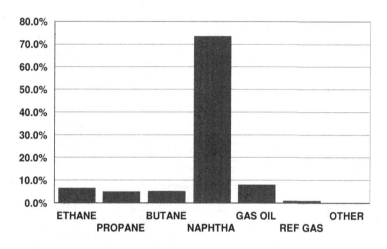

Figure 1.4: E.U. ethylene feedstock (2008)

The individual cracking operations across the countries of the EU and their nameplate capacities are shown in Table 1.3.

Russia

Russia has an annual nameplate ethylene production capacity of 3.49 million tonnes. The plants are scattered across Russia from European Russia to the Far East (Table 1.4). By world standards most plants are small with capacities of 400 kt/y or less. A cross section of feedstocks is used.

Table 1.3 European Cracking Operations (Country, Capacity (t/y), Company, Location)

COUNTRY	TOTAL	COMPANY	LOCATIION	t/y
Austria	500000	OMV AG	Schwechat	500000
Belgium	2460000	BASF Antwerpen NV	Antwerp	1080000
		Benelux FAO	Antwerp	230000
		Benelux FAO	Antwerp	580000
		Benelux FAO	Antwerp	570000
Czech Rep.	485000	Chemopetrol AS	Litvinov	485000
Finland	330000	Borealis	Porvoo	330000
France	3373000	A.P. Feyzin	Feyzin	250000
		ExxonMobil	Notre Dame de Gravenchon	400000
		Naphthachemie	Lavera	740000
		Polimeri Europa France SNC	Dunkerque	370000
		Soc. Du Craqueur de L'Aubette	Berre L'Etang	450000
		Total Petrochemicals	Carling- St. Avoid	568000
		Total Petrochemicals	Gonfreville L'Orcher	520000
		Total Petrochemicals	Lacq	75000
Germany	5757000	Bassell Polyfine GMBH	Wesseling	738000
		Bassell Polyfine GMBH	Wesseling	305000
		BASF AG	Ludwigshafen	620000
		BP	Geisenkirchen	580000
		BP	Geisenkirchen	480000
		INEOS	Dormagen	550000
		INEOS	Dormagen	544000
		LyondellBassell	Munchmunster	320000
		Dow Chemical AG	Bohlen	560000
		OMV Deutschland GMBH	Burghausen	450000
		Shell DEA Mineraloel AG	Heide	110000
		Shell DEA Mineraloel AG	Wesseling	500000
Greece	20000	EKO Chemicals	Thessalonika	20000
Hungary	660000	Tiszai Vegyi Kombinat	Tiszaujvaros	370000
		Tiszai Vegyi Kombinat	Tiszaujvaros	290000
Italy	2170000	Polimeri Europa	Brindisi	440000
		Polimeri Europa	Gela	245000
		Polimeri Europa	Porto Marghera	490000
		Polimeri Europa	Priolo	745000
		Syndial	Porto Torres	250000

Table 1.3 (continued)

Netherlands	3975000	Dow Chemical Europe	Terneuzen	580000
		Dow Chemical Europe	Terneuzen	585000
		Dow Chemical Europe	Terneuzen	635000
		SABIC Europetrochemicals	Geleen	600000
		SABIC Europetrochemicals	Geleen	675000
		Shell Nederland Chemie	Moerdijk	900000
Norway	550000	Noretyl AS	Rafnes, Bamble	550000
Poland	700000	PKN Orlen	Plock	700000
Portugal	330000	Borealis	Sines	330000
Slovakia	200000	Slovnaft Joint Stock Co.	Bratislava	200000
Spain	1430000	Dow Chemical	Tarragona	580000
		Repsol Petroleo SA	Puertollano	250000
		Repsol Petroleo SA	Tarragona	600000
Sweden	625000	Borealis	Stenungrund	625000
UK	2855000	INEOS	Grangemouth	730000
		INEOS	Grangemouth	340000
		ExxonMobil Chemical CO.	Fawley	120000
		ExxonMobil Chemical CO.	Mossmorran, Fife	800000
		Huntsman	Wilton	865000

Table 1.4: Russian Petrochemical Operations

COMPANY	LOCATION	t/y
Angarskneftorgsintez	Angarsk, Siberia	60000
Angarskneftorgsintez	Angarsk, Siberia	240000
Nizhnekamskneftekhim	Nizhnekamsk	450000
Norsy	Norsy	300000
Omskykauchuyk	Omsk, Siberia	90000
Orgsintez	Kazan	140000
Orgsintez	Kazan	100000
Orgsintez	Kazan	100000
Orgsintez	Orsk	45000
Polimir	Novopolotsk	150000
Salavatneftorgsintez	Salavat	300000
Sibur Himprom	Perm	30000
Sibur Neftechim	Nizhniy Novgorod	300000
Sintezkauchuk	Samara	300000
Stavripolpolymer	Prikumsk	350000
Tomsk PCC	Tomsk	300000
Uraorgsintes	Ufa	235000

Ethylene Production in the Middle East[2]

Over the past decade (to 2008) there has been an enormous expansion in the production of olefins and resins in the Middle East. This has been driven by: (i) the availability of feedstock at low prices as a consequence of the large oil reserves and (ii) the strategic location of the Middle East in being able to supply both the Atlantic and Far East petrochemical demand - in particular the enormous rise in demand from China.

As of 2008 installed capacity based on ethylene is 10.4 million tonnes across five nations of the Middle East as illustrated in Figure 1.5.

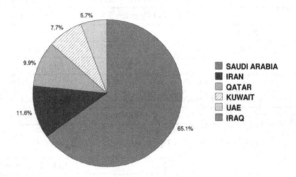

Figure 1.5: Middle East Ethylene Capacity (10.4 million tonnes 2008)

Most capacity is in Saudi Arabia, which has almost 7 million tonnes of ethylene capacity mainly using gas based feedstock[3]. Current capacity in Iran, Qatar and Kuwait stands at about 1 million tonnes each and the UAE has a cracker of 600,000 tonnes. Larger plants are under construction in Iran[4]. The status of the Iraqi petrochemical industry is unknown.

The feedstock used in the Middle East is illustrated in Figure 1.6. What distinguishes cracking operations in the Middle East from those of other regions is the dominance of ethane cracking over other feedstocks.

As is illustrated in the Figure 1.6, ethane is the major feedstock of the region. Along with propane and butane, ethane is extracted from natural gas either as gas associated with oil or from large natural gas fields developed for LNG production, as in Qatar.

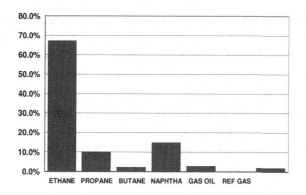

Figure 1.6: Middle East feedstock slate (2008)

A critical factor underpinning the growth in petrochemicals is that the Middle Eastern governments have made the political decision that the pricing of gas and its derivatives (ethane, propane and butane) is not related to the price of crude oil for domestic petrochemical purposes. This distinguishes the Middle East operations from many of those elsewhere such as Europe and USA where the pricing of gas derived feedstock bear a relationship with the prevailing price of crude oil.

When oil price was low (as in the mid 1990s) feedstock costs where broadly similar across the world. However, in a time of high gas and oil prices, the Middle East pricing regime has introduced a large feedstock differential in their favour. This arrangement delivers an enormous competitive advantage to Middle East producers of ethylene which use gas based feedstock. This flows through to the production costs of ethylene derivatives such as polyethylene resins, ethylene glycol, ethanol and acetic acid. The issue of differential feedstock pricing is of increasing concern to the EU where a large part of the downstream products are sold and where domestic EU producers pay much higher prices for feedstock. The issue is a point of contention in the trade between the Middle East and the EU.

Also important in future developments are those cracking operations based on feedstock from the large Qatar North Gas Field[5]. Further gas based plants are planned in Saudi Arabia to come on-stream from 2008. However, one short term issue will be the viability of Iran's

Petrochemical Economics

industry if the current imbroglio over uranium enrichment results in trade sanctions.

Should these projects come to fruition, the Middle East producers will be the lowest cost producers for a wide range of petrochemicals and derivatives. The major portion of the products would be exported to the world markets and so will impact on the world price. This will be a particular concern to most producers in Europe and the Far East with feedstock (naphtha) linked to the prevailing crude oil price. The cracking operations in the Middle East in 2008 are listed in Table 1.5.

Table 1.5: Ethylene Producers in the Middle East (2008)

COUNTRY	COMPANY	LOCATION	t/y
Iran	Amir Kaibar Petrochemical Co.	Amir Kabir	520000
	Arak Petrochemical	Arak	247000
	Bandar Imam Petrochemical	Bandar Imam	311000
Kuwait	Equate Petrochemical	Shuaba	800000
Qatar	Qatar Petrochem. Co.	Mesaieed	530000
Saudi Arabia	Al Jubail Petrochemical Co.	Al Jubail	800000
	Arabian Petrochemical	Al Jubail	650000
	Arabian Petrochemical	Al Jubail	800000
	Arabian Petrochemical	Al Jubail	800000
	Al Jubail Petrochemical Co.	Al Jubail	1000000
	Saudi Petrochemical Co.	Al Jubail	1045000
	Yanbu Petrochemical Co.	Yanbu	875000
UAE	Bourouge Abu Dhabi Polymers	Ruwais, Abu Dhabi	600000

One aspect of the developments is that many of the producer organisations have access to the latest technologies. One company, SABIC, owns major petrochemical plants in the EU and has now a strong research and development arm producing new technologies and product improvements.

Cracking Operations in the Far East[6]

The nameplate capacity of ethylene plants in the Far East is now over 28 million tonnes. This corresponds to over 25% of the world's total ethylene capacity. The countries contributing to this total are shown in Figure 1.7.

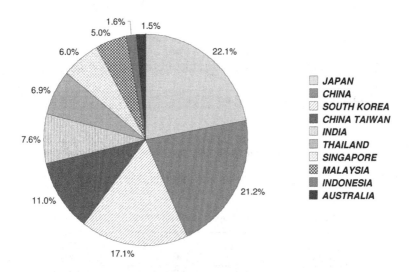

Figure 1.7: Far East Ethylene Capacity (32.9 million tonnes 2008)

Japan remains the country with the highest nameplate capacity with over 22% of the regions total. China and South Korea follow this with about 21% and 17% respectively. Taiwan (11%) and India (8%), Singapore (6%), Malaysia (5%) and Thailand (7%) are significant players in the region. Indonesia and Australia, with their production remaining static, remain outsiders to this growth in the region's ethylene production capacity (each below 2% of the region's production capacity).

Over the past decade, the ethylene capacity in the Far East has grown on average of 9% each year. This is considerably higher than the world growth rate of 5% over the same period. This growth to 2008 is illustrated in Figure 1.8.

Most of this growth has concerned the growth of China and suppliers of commodity resins and chemical intermediates to the rapidly growing Chinese market.

Table 1.6 lists the 2008 nameplate capacities in the Far East by country and the average annual growth over the previous decade.

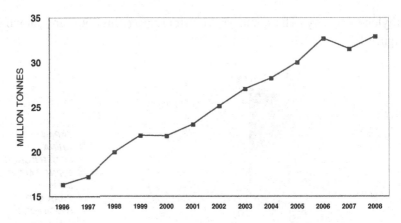

Figure 1.8: Growth in Far East Ethylene Capacity

Table 1.6: Nameplate Capacities for Ethylene Cracking (2008)

COUNTRY	t/y	growth/y
Australia	502000	-0.1%
China	6988000	8.9%
China Taiwan	3621000	25.7%
India	2515000	8.1%
Indonesia	520000	-0.5%
Japan	7265000	-0.1%
Malaysia	1649000	20.0%
Singapore	1980000	34.0%
South Korea	5630000	5.3%
Thailand	2272000	20.3%
TOTAL	32942000	6.5%

Comparing the growth rates we see that there has been a steady annual growth rate in the Far East petrochemical capacity in India and China. Most of the regions growth has been in Taiwan and in the South East Asian nations of Malaysia, Singapore and Thailand which have seen annual average growth rates over 20%. The nameplate capacities of these countries far outstrip local demand. These are export industries which supply the growing markets in India and in particular China whose industries, despite an 8% growth rate, have failed to keep up with rising demand.

Over the decade there has been some increase in capacity in South Korea (5.3%) which is close to the regions average growth rate of 6.5%. Australia, Indonesia and Japan have not changed significantly in capacity, the economies of these nations relying more on imports from South East Asia and the Middle East.

Feedstocks

In the Far East naphtha remains the dominant feedstock. Ethane is used in several countries, where it is available from local natural gas developments. There has been a continued decline in the use of gas oil. LPG is a minor contributor to feedstock in the region. The principal feedstocks used in the Far East are shown in Figure 1.9.

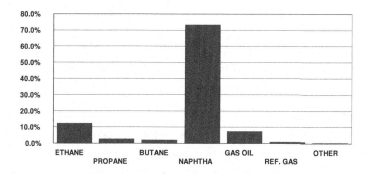

Figure 1.9: Feedstock breakdown in the Far East

However, there is a considerable variation between countries with some almost entirely naphtha and some substantially based on ethane and LPG. The various national breakdowns are shown in Figure 1.10.

Far East Country Survey

Australia (502 kt/y)

There are two major centres one based on Botany Bay near Sydney and the other at Altona in Melbourne. They produce the bulk of the ethylene which is made from ethane, with some supplementary LPG at Altona and naphtha at Botany. A small ethane cracking operation (32 kt/y) at Footscray (Melbourne) produces ethylene for styrene manufacture.

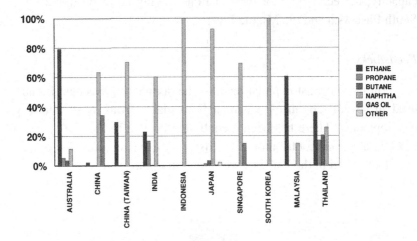

Figure 1.10: Feedstock breakdown by country in the Far East

China (6,9880 kt/y)[7]

The massive expansion of the Chinese economy has had a profound impact on the production and use of commodity plastics in China. The first is the considerable growth in demand for polymer products which have outstripped local supply and as a consequence China is a major importer. This has had the effect of promoting large export oriented plants in other Asian countries, with large parts of their product slate destined for the Chinese market. China is now second to Japan in the amount of ethylene produced in the Far East. The Chinese nameplate capacity is almost 7 million tonnes/year.

The plant locations and capacity are listed in Table 1.7. Most of the Chinese plants are old, with capacities below 200,000 t/y. Many of these plants were designed to use gas-oil and naphtha as feedstock. This takes advantage of some of China's indigenous crude oil, which have high levels of paraffin wax in the gas-oil fractions. In steam cracking, such gas-oils give high yields of ethylene and propylene. Newer plants are larger and tend to be more naphtha based. One plant (Panjin) uses ethane as feedstock.

The large growth in demand has spurred a large number of proposals to increase indigenous capacity. Many of these proposals have fallen by the wayside. Nevertheless, we would expect to see increasing capacity coming on-line over the next decade.

Table 1.7: Chinese Petrochemical Operations

COMPANY	LOCATION	t/y
BASF-YPC Co Ltd.	Nanjing	600000
CNOOC	Daya Bay, Guangdong	800000
CNOOC	Dushanzi	140000
China Petrochem. Industrial	Daqing	320000
Dalian Pet. Chem.	Dalian	4000
Fushan Pet. Chem Cpx.	Fushan	115000
Gaoqiao Petrochem	Gaoqiao	14000
Guangzhou Petrochem	Guangzhou	150000
Jilin Chemical	Jilin	700000
Langzhou Chemical Industrial	Langzhou	240000
Panjin Gas Processing	Panjin	130000
SINOPEC	Beijing	660000
SINOPEC	Guangzhou	140000
SINOPEC	Maoming	380000
SINOPEC	Neijing	650000
SINOPEC	Puyang Henan	180000
SINOPEC	Qilu	720000
SINOPEC	Shanghai #1	145000
SINOPEC	Shanghai #2	700000
SINOPEC	Tianjin	200000

China Taiwan (3,620 kt/y)

China Taiwan has a nameplate ethylene capacity of 3.6 million tonnes a year of ethylene. This makes Taiwan the fourth largest producer of olefins in the Far East. All of the production is from naphtha so that large volumes of propylene, higher olefins and aromatics are also produced. These feedstocks are used to produce a range of polymers, fibre intermediates and petrochemicals in large integrated complexes.

There are only two major players producing olefin monomers: The China Petrochemical Development Corporation (often referred to as CPC, but this can lead to confusion with other Taiwan and mainland Chinese organisations) was the original government owned organisation (now privatised) charged with development of Taiwan's petrochemical industry. It can be regarded as a diversified conglomerate with interest in housing and construction as well as chemicals. The Formosa Petrochemical Corporation (FPC) is a subsidiary of the Formosa Plastics Corporation and has major investments in ethylene and downstream petrochemical operations. The location and size of the current (2008) cracking operations are given in Table 1.8.

Table 1.8: Taiwan Petrochemical Operations

COMPANY	LOCATION	t/y
Chinese Petroleum Corp.	Kaohsiung Linyuan	422000
Chinese Petroleum Corp.	Linyuan	230000
Chinese Petroleum Corp.	Linyuan	419000
Formosa Petrochem. Corp.	Mailiao	450000
Formosa Petrochem. Corp.	Mailiao	900000
Formosa Petrochem. Corp.	Mailiao	1200000

All of the cracking operations use naphtha as feedstock. The future developments of Taiwan's petrochemicals business are linked to developments in the refinery sector which provide the feedstock. Until recently, the supply of petroleum products was in the hands of the government owned Chinese Petroleum Corporation (another CPC!) which operated three refineries and a condensate splitter. These operations supplied the downstream petrochemical plants with naphtha. However, the advent of Taiwan's entry into the WTO has broken this monopoly and FPC has established itself as Taiwan's first private refiner. The FPC refinery was built with the intention of facilitation feed supply to its new crackers. This opening of the petroleum market is allowing the entry of other players and some of the world oil majors have begun to enter the market.

As well as having some of the world's large integrated facilities, Taiwan has major world players in the downstream products and markets. Some of these operations are wholly owned subsidiaries of the major ethylene producing companies. However, there are some major independents that started out in a niche downstream market and have now expanded to become major players in the Far East chemicals industry.

Because many of the companies started in the downstream sector they have been open to further downstream high-tech sector investments. Furthermore, relative to mainland China, Taiwan's labour cost are high and this has seen a slowing of investment in traditional petrochemical operations in favour of placing such investment in the mainland. However, there is a major shift underway into the higher added value engineering plastics, electronic plastics (LCDs) and biotechnology and biomaterials. These future developments in advanced materials are being spurred by Japanese investment in joint ventures.

India (2.515 kt/y)[8]

The past ten years has seen a spectacular growth in the Indian petrochemicals and polymer industries so that today India is a major player in the region. India now has a nominal ethylene capacity of 2.5 million tonnes of ethylene, which places it fifth in terms of capacity in the Far East.

The per capita consumption of polymers at 2.7 kg/y is way below the world average (c. 19 kg/y; compare the developed world >70 kg/y). The demand is growing rapidly at about 12% per annum. This means that most of the new capacity is focused on the increasing domestic demand rather than in the production of export product (compare with Singapore with its export oriented industry). However, in the short term there may be some over-capacity and hence the potential for exports. Most of the plants are on the western side of India. The principal ethylene producing facilities are listed in Table 1.9.

Reliance Industries Ltd (RIL) is a large industrial conglomerate with interests in petrochemicals, refining, textiles, power generation, oil and gas exploration and telecommunications. RIL is one of the world's major manufacturers of plastics and polymers. It has over 50% market share of the Indian market and claims to be the 6th largest PP producer in the world with a capacity of 400,000 t/y. It has a large ethylene cracker (750,000 t/y, naphtha feed) at Hazira and is the majority owner of a very large refinery at Jamnagar, both in Gujarat. The refinery produces large volumes of propylene for PP production.

Table 1.9: Indian Petrochemical Operations

COMPANY	LOCATION	t/y
Gas Authority India	Pata, Utta Pradesh	300000
Haldia Petrochemicals	Haldia, West Bengal	520000
Indian Petrochemicals Corp.	Baroda, Gujarat	130000
Indian Petrochemicals Corp.	Gandhar, Gujarat	300000
Indian Petrochemicals Corp.	Nagothane	400000
National Organic Chemical Ind.	Thane, Maharashtra	75000
Reliance Industries	Hazira, Gujarat	790000

Petrochemical operations are vertically integrated through polyester and fibre intermediates to large textile operations. It is one of the world's largest producers of *para*-xylene and PTA.

Indian Petrochemical Corporation Ltd. (IPCL) is a government owned corporation with the remit to expand the petrochemical and plastics production of India. It has a naphtha based ethylene cracker at Vadodara (132,000 t/y), and large gas based complexes at Nagothane near Mumbai (Bombay; 400,000 t/y) and Dahej near Bharuch in Gujarat (300,000 t/y). The company produces polymers, fibre intermediates, catalysts and absorbents.

Gas Authority of India (GAIL) is a government authority that markets gas produced by various upstream producing consortia. GAIL operates a 300,000 t/y ethylene cracker at Pata in Uttar Pradesh. The ethylene is processed downstream to HDPE and LLDPE.

Haldia Petrochemicals Ltd (HPL) has a large naphtha cracker at Haldia in eastern India (West Bengal). Downstream the company operates LLDPE, HDPE and PP plants.

National Organic Chemical Ind. Ltd (NOCIL) was established in the early 1960s with a series of collaborative agreements with Shell and UOP and was the first company to set up a naphtha based cracker in

India at Thane near Mumbai (Bombay). The plant is small by today's standards with a capacity of 75,000 t/y ethylene. NOCIL produces petrochemicals and rubbers.

Oswal Agro Mills Ltd. is an agricultural company with several fertilizer plants with a small (naphtha) cracker producing 22,000 t/y ethylene near Mumbai (Bombay). Chemplast Sanmar Ltd. (CSL) is a small company focusing on the production of chloro-chemicals. Based in Tamil Nadu, the company has a small ethylene plant which uses ethanol as a feedstock.

Indonesia (520 kt/y)

Indonesia is currently a minor player in the Far East olefins industry. The industry is centred on a single world-scale naphtha cracker at Cilegon in West Java. PT Chandra Asri owns the plant. The feedstock is entirely naphtha. As built, the plant has an ethylene capacity of 515,000 t/y of ethylene and 240,000 t/y propylene which feeds several downstream operations.

Japan (7,265 kt/y)[9]

Japan has a major slice of the ethylene production capacity in the Far East, with nameplate capacity of approximately 23% of the area's total nameplate capacity. Japan has a ethylene production capacity of about 7 million tonnes per year. This nameplate capacity has been stable since the mid-1990s and growth is expected to be modest, mainly by de-bottlenecking operations. This static growth in capacity is in contrast to most of the other countries in the Far East which have seen large increases in capacity since the latter part of the 1990s. This has resulted in Japan's share of capacity in the Far East falling from 41% in 1995 to about 23% today.

The production capacity is in the hands of 10 manufacturing companies. The names, locations and capacities are shown in Table 1.10. Four companies [Mitsubishi, Mitsui (through Ukishima and Keiyo Ethylene), Idemetsu and Showa Denko] hold 68% of the nameplate capacity.

Table 1.10: Japanese Petrochemical Operations

COMPANY	LOCATION	t/y
Asahi Kasei Corp	Kurasiki, Okayama	484000
Idemetsu Petrochem.	Chiba	374000
Idemetsu Petrochem.	Tokuyama	450000
Keiyo Ethylene	Ichihara, Chiba	768000
Maruzen Petrochemicals	Chiba	480000
Mitsubishi Chemical Corp.	Kashima	375000
Mitsubishi Chemical Corp.	Kashima	453000
Mitsubishi Chemical Corp.	Mizushima	450000
Mitsui Chemicals Inc.	Chiba	553000
Mitsui Chemicals Inc	Takaishi City, Osaka	450000
Nippon Petrochemical	Kawasaki	450000
Showa Denko	Oita	600000
Sumitomo Chemical Co. Ltd.	Chiba	380000
Tonen Chemical Corp.	Kawasaki	505000
Tosoh Corp.	Yokkaichi, Mie	493000

Apart from one plant of Keiyo Ethylene (a subsidiary of Mitsui) and one plant of Mitsubishi, all of the producing plants are over 10 years old, most are over 25 years old. This means that most of the capital is fully depreciated and most plants can operate on a basis ignoring capital costs. This helps the Japanese operations to survive periods of depressed ethylene prices.

Of the total ethylene production about 68% is used immediately near the plant by subsidiary companies and affiliates. About 28% is sold on the merchant market and about 4% is exported.

The large merchant trade (about 2 million tonnes per year) is helped by an extensive pipeline system with 88% of ethylene being delivered by pipeline to the end user. The remaining 12% (about 800,000 t/y) is delivered by ship or barge, to the largely coastal petrochemical plants in Japan. The fleet dedicated to intra-Japan trade comprises about 11, mostly refrigerated, vessels with a range of capacities from 440 tonnes to 1700 tonnes of ethylene. Shipping terminals for ethylene facilitate a small import trade in ethylene of about 20,000 t/y.

The dominant feedstock is naphtha, although in some cases hydrogenated natural gas liquids (H-NGL or condensates) are used. However, the choice of condensate is probably restricted to those with a low end point (i.e. they are very similar to naphtha such as A-180 from Saudi Arabia). There is a small use of LPG (butane and propane) in some of the cracking operations.

Like many countries in the Far East, there is a relatively high demand for propylene. To maximise propylene production from naphtha cracking, the process plant is operated at low severity. In order to maintain design levels of ethylene, more naphtha feedstock is required, with the naphtha requirement being about 3.8 times the weight of ethylene produced. This creates a large demand of about 750,000 to 800,000 bbl/d for petrochemical (paraffinic) naphtha.

Most naphtha (65%) is imported, the rest is produced domestically by distilling crude oil in refineries. Due to the large demand and concomitant international trade, it is the Japanese petrochemical market that sets the specification for traded naphtha in the Far East - the so-called "Japanese open spec.". Most producers of naphtha ensure that their product meets this specification as is illustrated in by the data in Table 1.11.

Table 1.11: Japanese Open Spec and Some Typical Naphtha Compositions

	B.P. (C)	DENSITY (kg/l)	PARAFFINS
Japan Open Specification	24 to 204	0.665 to 0.740	65 Min
Cooper (Australia)	full range	0.729	69.6
Udang (Indonesia)	32 to 191	0.7264	75.5
Khafji (Kuwait)	32 to 190	0.7201	73.4
A-180 (Yanbu, Saudi Arabia)	36 to 154	0.6689	93.8

Naphtha cracking provides about 4.3 million tonnes of propylene per year, which is out of a total demand for propylene in excess of 5.3 million tonnes per year. The difference (about 20%) is made up by propylene extracted from refinery off-gases, particularly FCC operations (used to produce gasoline from heavier feed stocks such as heavy gas-oil or residua).

Korea (5,630 kt/y)

South Korea is a major player in the Far East olefins and poly-olefins markets with 17% of the regions total ethylene capacity of 32 million tonnes /year.

The current (2008) total nameplate capacity of the South Korean petrochemical industry is 5.63 million tonnes. The major players, location and nameplate capacity (2008) are shown in Table 1.12.

Table 1.12: South Korean Petrochemical Industry

COMPANY	LOCATION	t/y
Honam Peterochemical	Yeochun	700000
LG Daesan Petrochemical	Daesan	450000
Lotte Daesan Petrochemical	Daesan	600000
Korea Petrochem Ind.	Ulsan	320000
LG Petrochemical Co.	Yeosu City	760000
Samsung General Chemicals	Daesan	820000
SK Corp.	Ulsan	185000
SK Corp.	Ulsan	545000
Yeochon	Yeochun	420000
Yeochon	Yeochun	480000
Yeochon	Yeochun	350000

All of the plants use naphtha as feed and so produce a broad range of olefins and by-products enabling the production of a large range of products in large integrated complexes. The domestic demand is less than 50% of the production that is the petrochemical operations are export oriented. One aspect of the reliance of the Korean petrochemical sector on exports is the suspicion that during the depths of the petrochemical business cycles, the plants operate on a cash-cost basis. This allows them to undercut rivals having to service debt.

The financial crisis in the Far East during the late 1990s exposed the high debt levels of the petrochemical operations, which were not being serviced. This has forced restructuring of the industry in order to reduce debt levels. For instance some companies had debt/equity ratios of well over 300%. Since restructuring, these levels have been reduced, but are still typically in the 200% region. However, there has been some criticism of the restructure as involving too much financial engineering with total debt still similar to 1997 levels. How these operations will fare in the current crisis of 2009 is moot.

North Korea

As a consequence of the recent political events on the Korean peninsula, there is increasing interest in how the North Korean economy can be integrated into the economies in the Far East. North Korea has permitted some foreign investment in recent years and North Korea has recently asked the World Bank for guidance in establishing a market economy.

North Korea has a nominal ethylene capacity of 60,000 t/y at a plant in Pyongyang. This very small operation could expand should oil be discovered in offshore blocks currently being explored by western companies including Australia's Beach Petroleum.

A more promising basis for the development of chemical and petrochemical plants in the north might come as a consequence of any trans-Korean gas-pipeline developments from the very large Russian Kovylta gas fields at Irkutsk. This might provide both energy and feedstock (ethane) for future petrochemical developments.

Malaysia (1,649 kt/y)[10]

Although currently a minor player on the Far East petrochemicals scene, Malaysia has a strong and growing petrochemical sector with a nameplate ethylene capacity approaching 1.7 million tonnes per year. Led by Petronas (the national oil company), Malaysia has attracted over US$ 7.6 billion since 1997 and a further US$ 5 billion is committed from 2001.

There are three major ethylene plants that feed downstream operations. All are based on the Malaya Peninsula: Table 1.13.

The oldest plant and largest integrated petrochemical plant is at Kertih in Terengganu State. This complex uses gas from the major oil and gas fields off the eastern cost of the Peninsula.

Table 1.13: Malaysian Petrochemical Operations

COMPANY	LOCATION	t/y
Ethylene Malaysia	Kerith	400000
Optimal Olefins	Kerith	600000
Titan Petrochemicals	Pasir Gudang, Jahor	249000
Titan Petrochemicals	Pasir Gudang, Jahor	400000

The other major olefins plants are at Pasir Gudang in Jahor operated by the Titan Group. These plants utilise naphtha or LPG as feedstock that can be imported via the large Jahor port. Initially built around providing feed to poly-olefins plants, these facilities have expanded to produce aromatics. Clearly there is the potential for these developments to offer synergy with the large complexes in Singapore. Many of the downstream operations involve multinational corporations in a leading role.

Other chemicals operations are in Sarawak. Offshore gas feeds a large methanol plant (660 kt/y) on Labuan Island and an ammonia plant at Bintulu. Also at Bintulu is the large Shell Gas to Liquids plant, which produces high valued linear-paraffins and wax as by-products. The naphtha fraction from the GTL plant is used as petrochemical naphtha.

The petrochemical complexes in Malaysia are export driven. The competitive advantages lie in low priced gas feedstock and large integrated plants based on naphtha. The resulting complexes are able to deliver chemical intermediates throughout the Far East.

Singapore (1,980 kt/y)

The petrochemical operations in Singapore are based on Jurong Island. From a cluster of small islands in 1995, the site has been transformed by massive civil engineering to create a large flat land base dedicated to the production of petrochemicals and the integrated downstream industries. These infrastructure works alone have cost the Singapore government in excess of US$ 6,000 million to date. These developments are continuing and Singapore continues to attract private investment lured by the benefits of manufacturing chemicals on a world-scale fully integrated site in the Far East. The petrochemical operations are based around two major naphtha cracking operations. ExxonMobil Singapore (capacity 900 kt/y ethylene) is now complete and operational and Petrochemical Corporation of Singapore (PCS) has recently been expanded to over a million tonnes of ethylene. Another world-scale cracker is reported to be under consideration by Shell.

Cracker feedstock for the Island is entirely imported. Two large oil refineries (ExxonMobil Singapore Pte. Ltd. (227,000bbl/d) and Singapore Refining Company (285,000bbl/d)) supply naphtha to the main cracking operations and additional feedstock supply can be obtained from other Singapore refineries (Shell Eastern Petroleum (405,000bbl/d) on Pulau Bukom and ExxonMobil Oil Singapore (255,000bbl/d) on the mainland near Jurong). Undersea pipelines integrate all these facilities.

Juxtaposed to these main facilities are clustered a large number of chemical processing companies producing intermediates and finished petrochemical products. In order to achieve the greatest benefits, an integrated site requires the sharing of utility services. This minimises the capital requirements for investment by eliminating the need for power, steam, gas, shipping terminals etc. required for stand-alone facilities. Jurong Island's integration is achieved by the existence of a series of service industries dedicated to providing supporting services and utilities to the chemical plants.

Thailand (2,272 kt/y)[11]

There are four main olefins plants at Map Ta Phut in Rayong Province just south of Bangkok. These plants have a capacity of over 2 million tonnes of ethylene (Table 1.14), making Thailand a major player in Far East petrochemicals.

Table 1.14: Thai Petrochemical Industry

COMPANY	LOCATION	t/y	FEEDSTOCK
PTT Chemical	Map Ta Phut	437000	Ethane LPG
Rayong Olefins Co. Ltd..	Map Ta Phut	800000	LPG naphtha
PTT Chemical	Map Ta Phut	350000	Ethane
PTT Chemical	Map Ta Phut	385000	LPG naphtha
PTT Chemical	Map Ta Phut	300000	Ethane

South America

Although South America is a smaller player in the world petrochemical industry, three countries have significant and growing operations. The largest is Brazil (3.5 million tonnes) in six world scale operations. Feedstock for five of these is naphtha with the other based on ethane and LPG. Argentina has a nameplate capacity of 838 kt/y. Three of these plants are small local operations. Venezuela has a nameplate capacity of 600 kt/y in two operations. The operations, locations and feedstock are detailed in Table 1.15.

Table 1.15: Some South American Petrochemical Operations

COUNTRY	COMPANY	LOCATION	t/y	FEEDSTOCK
Brazil	Braskem SA	Camacari Bahia	600000	Naphtha
	Braskem SA	Camacari Bahia	680000	Ethane LPG
	Copesul	Triunfo, RS	700000	Naphtha
	Copesul	Triunfo, RS	500000	Naphtha
	Petroquimica Uniao SA	Santo Andre, SP	500000	Naphtha
	Rio Polimeros	Duques De Caxais	520000	Naphtha
Argentina	Dow Chemical	Bahia Blanca	275000	Ethane
	Dow Chemical	Bahia Blanca	490000	Ethane
	Huntsman Corp.	San Lorenzo	21000	Propane/ Naphtha
	Petrobas Energia	Puerto San Martin	32500	Propane
	Petrobas Energia	San Lorenzo	20000	Propane
Venezuela	Pequiven – Petrochima	El Tablazo, Zulia	250000	Ethane Propane
	Pequiven – Petrochima	El Tablazo, Zulia	350000	Ethane

Africa

There are only a small number of cracking operations in Africa. The main producers are Egypt, Libya and Nigeria each with a capacity of about 300,000 t/y and South Africa with a capacity of 585,000 t/y. The latter production is integrated with the large coal and gas to liquids operations of Sasol.

Feedstock and Carbon Emissions

Based on nameplate capacity, the relative amounts of feedstock used are shown in Figure 1.11. This graph illustrates that the two largest feedstocks are ethane and naphtha with naphtha accounting for over 50% of the required feedstock. LPG (propane, butane) and gas oil make a contribution, but in total this is less than 20%.

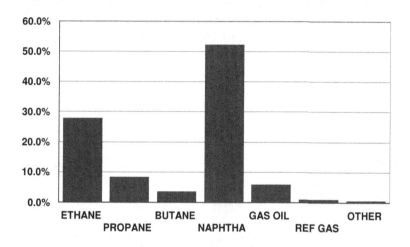

Figure 1.11: World ethylene feedstock

A typical naphtha cracking operation will use approximately 3.3 tonne naphtha per tonne of ethylene. Using this as a basis, the world demand for petrochemical naphtha is almost 200 million tonnes per year or almost 5 million barrels of naphtha per day. The ethane required is typically 1.3 tonnes of ethane per tonne of ethylene. This translates into 41 million tonnes of ethane per year. Most of this is derived from natural gas which (on a weight basis) contains about 10% ethane, hence some 400 million tonnes of natural gas is required to be processed to provide the world's petrochemical ethane or about 63 bcf/d of raw natural gas.

Ethylene cracking operations produce carbon dioxide emissions from fuel oil consumed in furnace operations and losses as a consequence of operational issues (flaring). Using the above data, the estimate of the world's emissions is 255 million tonnes of carbon dioxide. The breakdown by feedstock is shown in Figure 1.12.

Ethane and LPG cracking give little product other than ethylene and propylene. However, naphtha and gas oil produce large quantities of by-products such as pyrolysis gasoline. Assigning some of the carbon dioxide produced to these by-products lowers the carbon dioxide emission attributable to the olefins. Although naphtha produces much higher levels of carbon dioxide than ethane, distributing the emission over all the products produced lowers the total emission from naphtha to appoint where it its lower than for ethane cracking.

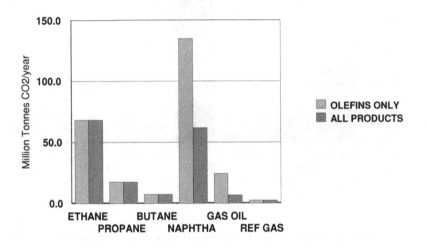

Figure 1.12: World carbon dioxide emissions from ethylene production

One of the issues facing the world petrochemical industry is the issue of placing some sort of emissions charge on carbon dioxide emitting industries, this is especially true for the developed economies which generally use naphtha feedstock. The application of a carbon emission charge would encourage the relocation and investment in many of the developing countries with emerging petrochemical industries. Many of these countries, especially in the Middle East, use ethane as the feedstock and as illustrated such a move may not necessarily result in lower overall emissions.

[1] *Oil & Gas Journal*, "International Survey of Ethylene from Steam Crackers", July 2008; see also W. Sedriks, *ibid.*, Nov. 5, 2001, p. 58

[2] A. W. Al-Sa'doun, *Oil & Gas Journal*, Nov. 13, 2000, p. 52

[3] A. M. Aitani, *Oil & Gas Journal*, Jul. 29, 2002; A.Al-Sa'doun, *ibid.*, Jan 2, 2006, p. 52 and Jan 9, 2006, p. 48

[4] M.H. Buffeboir, J.M. Aubury, X. Hurstel, , *Oil & Gas Journal*, Jan. 19, 2004, p. 60; A. Aik, S. Adibi, *ibid.* Mar. 26, 2007, p. 48

[5] T. Chang, *Oil & Gas Journal*, Aug 20, 2001, p. 72

[6] See also G. Kin, *Hydrocarbon Asia*, Jul/Aug 2006, p. 48

[7] Wang Yong, *Hydrocarbon Asia*, Sep. 2002, p.16

[8] J. W. King, *Oil & Gs Journal*, Feb. 11, 2002, p. 58

[9] Anon., *Hydrocarbon Asia*, Jul/Aug 2005, p. 16; *idem.*, Nov/Dec 2002, p. 30

[10] Anon., *Oil & Gas Journal*, Sep. 18, 2000, p. 58; Anon., *Hydrocarbon Asia*, Nov/Dec 2005, p. 8

[11] Anon., *Hydrocarbon Asia*, Sep/Oct. 2005, p. 10

CHAPTER 2

CHEMISTRY OF OLEFIN PRODUCTION

The principal olefins for the production of polymers and resins are ethylene and propylene. These are made by cracking larger molecules, which for the most part are paraffins. Two processes are involved – thermal cracking (pyrolysis) and catalytic cracking. Of these two types the former is the dominant process for the production of ethylene and propylene whilst the latter makes a significant contribution to the production of propylene.

The academic and patent literature of hydrocarbon pyrolysis is very large. An extensive exposition of various aspects of pyrolysis is given by Albright *et al.*[1] to which the reader is referred for greater detail of many aspects of the industrial uses of pyrolysis. This chapter gives the salient features of the chemistry of hydrocarbon pyrolysis as it applies to describing the key points of the technology and economics of production of olefins.

We are concerned with the breaking of carbon–carbon and carbon– hydrogen bonds in large molecules by thermal processes. These processes occur by the means of free radicals. It is the production of free radicals and the subsequent rearrangement which produces the products of the steam cracking plants. The free radical chemistry generally occurs in the gaseous or liquid phase away from surfaces, and is thus distinguished from catalytic pyrolysis which requires a usually acidic surface to proceed. The chemistry of catalytic cracking processes is important in the production of propylene in fluidized-bed cracking operations.

The distinguishing feature of thermal (free radical) cracking in the gaseous phase and acid catalysed processes is that the former leads to ethylene as a major product. Ethylene is only a minor product in catalytic processes and where it is present in catalytic processes it can be argued that this is a consequence of thermal processes. When surfaces are present in thermal processes, this tends to lead to unwanted formation of carbon or coke.

A key technical difference between the two approaches is that thermal cracking of hydrocarbons to ethylene is usually performed at temperatures in excess of 800°C, whereas catalytic processes occur generally below 550°C.

Thermodynamics of Thermal Cracking

For the most part we are concerned with the breaking of carbon-carbon and carbon-hydrogen bonds and subsequent rearrangement of intermediate free radicals to produce ethylene. These bonds are broken by the simple application of temperature and because the bonds of interest are strong, high temperatures are required.

Ethylene is produced in large quantities in many countries by the thermal pyrolysis of ethane with the generalised stoichiometry:

$$C_2H_6 = C_2H_4 + H_2$$

The key features of the thermodynamics of ethane pyrolysis are illustrated in Figure 2.1[2], which shows the free energy relationship of ethane to the product ethylene and other compounds of interest over a range of temperatures. This graph illustrates several points which are central to the technology and production economics of ethylene production:

Over most of the temperature range, all of the compounds have positive free energies. This means that they are unstable relative to the elements. Thus the most favoured thermodynamic products are carbon and hydrogen.

$$C_2H_6 = 2C + 3H_2$$

and

$$C_2H_4 = 2C + 2H_2$$

(kJ/carbon atom)

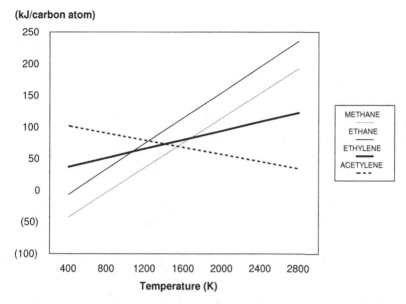

Temperature (K)

Figure 2.1: Thermodynamics of cracking – free energy of components

If the elements are the most favoured products, success in the production of ethylene will be achieved by preventing thermodynamic equilibrium occurring. This is accomplished by reducing the temperature of pyrolysis as quickly as possible (quenching) in order to prevent the products converting into the elements. In addition, surfaces promote the formation of carbon and there are several proprietary approaches to passivating the surfaces of the equipment in order to reduce carbon formation[3].

The lines for the free energy of formation of ethane and ethylene cross at about 1000K. Relative to ethane, ethylene becomes favoured at this and higher temperatures. The successful pyrolysis of ethane thus takes place at temperatures over 1000K.

When ethane is cracked to ethylene and hydrogen, there is a volume expansion and thus, by the Le Chatelier's Principle, the pyrolysis is favoured by lowering the pressure. In practice this is achieved by adding large volumes of steam so as to lower the partial pressure of the hydrocarbons. Steam addition also has the advantage of removing coke by steam gasification:

$$C + H_2O = CO + H_2$$

The Figure 2.1 also shows that up to about 1700K, methane is more thermodynamically favoured than ethylene and is hence a potential significant product of the pyrolysis process.

As temperatures rise acetylene becomes an increasingly favoured product. At temperatures higher than about 1400K, acetylene is more favoured thermodynamically than ethylene:

$$C_2H_6 = C_2H_2 + 2H_2$$

Minimising the pyrolysis temperature prevents the formation of acetylene. Conversely, if acetylene is a required product, production is maximised by high pyrolysis temperatures.

Most ethylene is produced by the pyrolysis cracking of heavier hydrocarbons such as naphtha. Figure 2.2 illustrates the key thermodynamic features using heptane as a proxy for the feedstock.

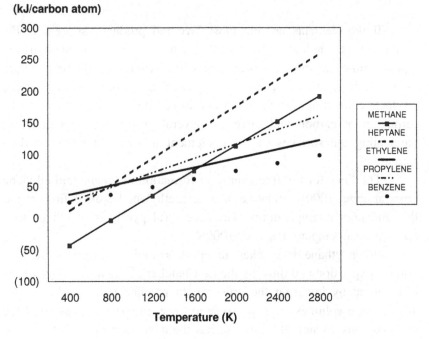

Figure 2.2: Hexane cracking – free energy of products

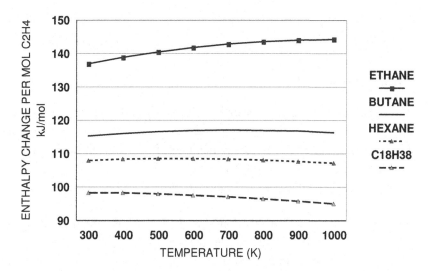

Figure 2.3: Enthalpy of paraffin cracking

In addition to the remarks made for the cracking of ethane, the graph illustrates the following points pertinent to cracking heavier molecules.

The cross over point of the free energy curves of heptane and ethylene occur at a lower temperature than that of ethane, namely at about 800K. Pyrolysis of naphtha to ethylene can therefore be practiced at lower temperatures than that required for the pyrolysis of ethane.

Over the temperature range of about 800–1200K, the free energy of ethylene and propylene are very similar and thus would be approximately equally favoured, lower temperatures in the range favours propylene over ethylene.

Benzene is more thermodynamically favoured than the two olefins and is expected to be a product of the pyrolysis.

Enthalpy of Conversion

The enthalpy of cracking reactions is shown in Figure 2.3. Ethane is seen to require the highest input of energy to produce ethylene with the enthalpy falling with increasing molecular weight of the feedstock. The implication is that higher furnace heat input is required for cracking lighter feedstock which in turn will lead to higher carbon dioxide emissions.

Chemical Kinetics[4]

The thermal pyrolysis of hydrocarbons proceeds by free radical chain reaction processes. These processes are exceedingly complex and this overview concentrates on the details as it impacts on the technology and economics of olefin production.

Radical chain processes comprise three reaction types.

(i)   Initiation reactions in which radicals are formed by the scission of carbon-carbon or carbon-hydrogen bonds. Reactions such as these involve the breaking of strong bonds and proceed by overcoming high activation energy at high temperatures. For example, the cracking of ethane to form two methyl radicals and the cracking of a hydrogen-carbon bond in ethane to form an ethyl radical and hydrogen radical:

$$C_2H_6 = 2CH_3^{\bullet}$$

and

$$C_2H_6 = C_2H_5^{\bullet} + H^{\bullet}$$

(ii)   Chain propagation reactions which continue the chain by transferring a radical from one moiety to another or rearrangement of a molecule and thereby generating an intermediate or product. For example, in ethane cracking reactions which produce methane and hydrogen and another radical and reactions which produce ethylene and another radical. These reactions occur at low to moderate activation energies and involve all possible combinations of intermediates and products. Because of the low activation energy and the high temperature required to initiate propagation these reactions are fast.

$$CH_3^{\bullet} + C_2H_6 = CH_4 + C_2H_5^{\bullet}$$

$$H^{\bullet} + C_2H_6 = C_2H_5^{\bullet} + H_2$$

$$C_2H_5^{\bullet} = C_2H_3^{\bullet} + H_2$$

$$C_2H_3^{\bullet} + C_2H_6 = C_2H_4 + C_2H_5^{\bullet}$$

(iii)   Chain termination reactions which eliminate radicals and thereby stop the process. These are typically radical combination reactions which occur at very low activation energies, hence are immediate when radicals meet, for example the reaction of a vinyl and a hydrogen radical to form ethylene.

$$C_2H_3^{\bullet} + H^{\bullet} = C_2H_4$$

Note that in radical chain processes, ethylene is produced by a series of reactions. The direct conversion of ethane to ethylene and hydrogen is absent from the scheme:

$$C_2H_6 = C_2H_4 + H_2$$

To illustrate the complexity of the process of thermal cracking, Table 2.1 lists some of the more important reactions in the cracking of propane.

This table illustrates that even for small molecules, the chain initiation, propagation and termination reactions are extensive. All possible products can be formed, but again the direct conversion reactions are absent, namely propane to propylene and hydrogen, or propane to ethylene and methane.

Another key point to note is that chain transfer and termination by radical combination leads to radicals and molecules with more carbon atoms than the feed (propane). Subsequent involvement of these moieties in the radical chain propagation leads to larger molecules. In practice this manner of radical cracking of ethane and propane cracking leads to some C_4, C_5 and C_{6+} products forming pyrolysis gasoline.

Table 2.1: Propane Cracking Reactions

INITIATION	$C_3H_8 = CH_3^{\cdot} + C_2H_5^{\cdot}$
PROPAGATION	$CH_3^{\cdot} + C_3H_8 = C_3H_7^{\cdot} + CH_4$
	$C_2H_5^{\cdot} + C_3H_8 = C_3H_7^{\cdot} + C_2H_6$
	$C_2H_5^{\cdot} = H^{\cdot} + C_2H_4$
	$CH_3^{\cdot} + H_2 = H^{\cdot} + CH_4$
	$H^{\cdot} + CH_4 = CH_3^{\cdot} + H_2$
	$C_2H_5^{\cdot} + H_2 = H^{\cdot} + C_2H_6$
	$C_3H_5^{\cdot} + C_2H_4 = CH_3^{\cdot} + C_3H_6$
	$H^{\cdot} + C_2H_4 = C_2H_5^{\cdot}$
	$C_2H_6 = 2CH_3^{\cdot}$
	$CH_3^{\cdot} + C_2H_6 = C_2H_5^{\cdot} + CH_4$
	$CH_3^{\cdot} + C_2H_4 = C_2H_7^{\cdot}$
	$C_3H_7^{\cdot} + C_2H_6 = C_2H_5^{\cdot} + C_3H_8$
	$H^{\cdot} + C_2H_6 = C_2H_5^{\cdot} + H_2$
	$H^{\cdot} + C_3H_6 = C_3H_7^{\cdot}$
	$C_3H_7^{\cdot} = CH_3^{\cdot} + C_2H_4$
	$C_3H_7^{\cdot} = H^{\cdot} + C_3H_6$
	$C_3H_7^{\cdot} + H_2 = H^{\cdot} + C_3H_8$
	$C_3H_7^{\cdot} + CH_4 = CH_3^{\cdot} + C_3H_8$
	$H^{\cdot} + C_3H_8 = C_3H_7 + H_2$
	$CH_3^{\cdot} + C_3H_6 = C_4H9^{\cdot}$
	$C_2H_5^{\cdot} + C_2H_4 = C_4H9^{\cdot}$
	$C_4H9^{\cdot} = H^{\cdot} + C_4H_8$
	$C_4H9^{\cdot} + H_2 = H^{\cdot} + C_4H_{10}$
	$C_4H9^{\cdot} = CH_3^{\cdot} + C_3H_6$
	$C_4H9^{\cdot} = C_2H_5 + C_2H_4$
	$C_2H_5 + C_4H_{10} = C_4H9^{\cdot} + C_2H_6$
	$C_4H_{10} = CH_3^{\cdot} + C_3H_7^{\cdot}$
	$C_4H_{10} = 2C_2H_5^{\cdot}$
	$H^{\cdot} + C_4H_{10} = C_4H9^{\cdot} + H_2$
	$CH_3^{\cdot} + C_4H_{10} = C_4H9^{\cdot} + CH_4$

Table 2.1 (continued)

TERMINATION	$2H^{\cdot} = H_2$
	$H^{\cdot} + CH_3^{\cdot} = CH_4$
	$H^{\cdot} + C_2H_5^{\cdot} = C_2H_6$
	$H^{\cdot} + C_3H_7^{\cdot} = C_3H_8$
	$2CH_3^{\cdot} = C_2H_6$
	$CH_3^{\cdot} + C_2H_5^{\cdot} = C_3H_8$
	$2C_2H_5^{\cdot} = C_2H_4 + C_2H_6$
	$2C_2H_5^{\cdot} = C_4H_{10}$
	$CH_3^{\cdot} + C_3H_7^{\cdot} = CH_4 + C_3H_6$

The formation of molecules larger in molecular weight than the feed is a feature of radical cracking processes. It is clearly demonstrated in the liquid phase thermal cracking of cetane ($C_{16}H_{34}$) where olefins with 32 carbon atoms are formed[5]. In fact, in this process some 10% of the resulting products have higher molecular weight than the feed. Continuing this process leads to coke precursors and coke, which as stated above is the thermodynamically favoured product.

Yield of Ethylene and Propylene from Lighter Feedstock

In many parts of the world ethylene and propylene are made from light gaseous feedstock – ethane, propane and butanes. These are often derived from large scale gas processing operations. Table 2.2 gives the typical single pass and ultimate (feed recycled to extinction) yields of products in steam-cracking processes. All of the products of significance are shown with ethylene and propylene in bold.

Ethane, as might be expected, shows the highest selectivity to ethylene. However, note that the pass conversion is low at typically 60%, with apparently 40% of the ethane feed passing through. This is because as indicated previously (Figure 2.1) ethane requires a high cracking temperature. The other products of note are hydrogen and methane which at 3-4% by weight in the stream occupy a large portion of the stream

Table 2.2: Typical Yields for Gaseous Feedstock

	Single Pass Yield				Ultimate Yield			
	C_2	C_3	n-C_4	iso-C_4	C_2	C_3	n-C_4	iso-C_4
Products								
Hydrogen	3.72	1.56	1.49	1.08	6.20	1.68	1.55	1.35
Methane	3.47	23.65	19.90	16.56	5.78	25.43	20.73	20.70
Acetylene	0.42	0.77	1.07	0.72	0.70	0.83	1.11	0.90
Ethylene	**48.82**	**41.42**	**40.59**	**5.65**	**81.37**	**44.54**	**42.28**	**7.06**
Ethane	40	3.48	3.82	0.88	0.00	3.74	3.98	1.10
allene/propyne	0.2	1.09	1.07	2.34	0.33	1.17	1.11	2.93
Propylene	**0.99**	**12.88**	**13.64**	**26.35**	**1.65**	**13.85**	**14.21**	**32.94**
Propane	0.03	7	0.48	0.38	0.05	0.00	0.50	0.48
Butadiene	1.33	2.82	4.13	1.49	2.22	3.03	4.30	1.86
Isobutene				19.60	0.00	0.00	0.00	24.50
n-butenes	0.25	0.89	1.92		0.42	0.96	2.00	0.00
Isobutane				20.00	0.00	0.00	0.00	0.00
n-butane			4.00		0.00	0.00	0.00	0.00
C_{5+} aliphatics	0.46	1.37	3.24	2.35	0.77	1.47	3.38	2.94
BTX	0.31	3.07	5.25	2.40	0.52	3.30	5.47	3.00
TOTAL	100	100	100.60	99.80	100.00	100.00	100.62	99.75
$C_{2=}$ & $C_{3=}$					83.02	58.39	56.49	40.00

volume. Note that even for ethane feed there are measurable quantities of propylene, butadiene, C_{5+} aliphatic hydrocarbons and BTX (benzene, toluene and xylene) produced.

Propane cracking produces a minor amount of propylene in pyrolysis cracking, the major product olefin being ethylene, with a commensurate high yield of methane. Pass conversion is much higher with only 7% of propane in the product stream. Higher molecular weight products are more prevalent with a significant amount of BTX (over 3%).

Normal-butane gives very similar yields of ethylene and propylene to propane cracking. Methane is lower and more butadiene, C_{5+} aliphatic hydrocarbons and BTX are produced.

Isobutane reverses the relative composition of the olefin products with the major olefin being propylene (over 30% ultimate yield), double the yield for propane or n-butane, with a commensurate high methane yield. A major product is isobutene. The ultimate ethylene yield is only about 7%.

Note that in the yield of the desired olefins on a weight basis are about 80% for ethane and less than 60% of the other feeds.

Of the gaseous feeds ethane, propane and n-butane are preferred for the production of ethylene and high isobutane content should be avoided. However, for some operations, propylene and isobutene are valuable products and butane streams of high isobutane content can be preferred[6].

Thermal Cracking of Larger Molecules

The cracking of naphtha produces most of the world's ethylene. Naphtha is the crude oil fraction boiling from about 32°C to 192°C. The composition of naphtha made from crude oil comprises four basic components: linear paraffins, branched paraffins, naphthenes (*cyclo*-paraffins) and aromatics. The relative amount of these in naphtha is dependent on the source crude oil and varies widely.

The cracking of these larger molecules is extremely complex, however some important generalisations can be made. These are illustrated in Figure 2.4 which illustrates what happens when the carbon-carbon bonds of various types of molecules are ruptured.

Linear Paraffins can break at any of the carbon-carbon bonds which leads to a relative large number of C_2 fragments. For example for hexane, cracking in the middle (position C in Figure 2.4) gives two C_3 moieties. At position B, two C_2 and two linear C_4 fragments result because statistically there are two positions. Cracking at position A similarly gives two C_1 and two linear C_5 fragments. If these fragments go on to produce products then the result is that two methane, two ethylene, two propylene, two butene and two pentene molecules. However, the linear C_4 and C_5 moieties have a good chance of further cracking reactions to produce more ethylene. Molecules like n-hexane have low octane number (research octane number (RON) is 19 for hexane) and are less useful to petroleum refiners for the production of motor gasoline. However, the high yield of C_2 fragments on cracking make linear paraffins attractive to petrochemical operations for producing ethylene.

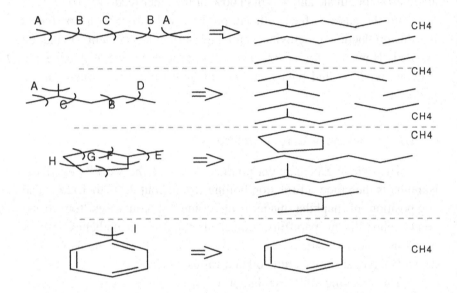

Figure 2.4: Fragmentation pathways for pertinent molecular types

Branched paraffins, as illustrated by 2-methylpentane, in a similar manner leads to three C_1 fragments, one C_2 fragment, two C_3 fragments, one branched C_4 and two linear C_5 fragments. Branched paraffins have relatively high octane (RON is 83 in the case of 3-methylpentane) and are attractive to refiners for the production of gasoline. The lower level of C_2 formation means they are less attractive to petrochemical operations.

Naphthenes, as illustrated by methyl-*cyclo*-pentane, leads to the formation of one C_1 fragment and a cyclo-C_5 fragment. The other possibilities are linear and branched C_6 fragments which can ultimately lead to C_2 and C_3 products. Naphthenes can have high octane (RON 107 for methyl-cyclo-pentane) and the potential high yield of C_2 and C_3 makes these molecules attractive to both refiners and petrochemical operators.

Aromatics, as illustrated by toluene, have very strong bonds in the ring. Furthermore, benzene is more thermodynamically stable than ethylene. The main reaction is cracking of the aromatic-aliphatic bonds to produce benzene and a C_1 fragment. These molecules are of little use to cracking operations but the high octane (RON 124 for toluene) makes them very attractive for gasoline production.

Olefins are not present in naphtha made from crude oil. However, some types of naphtha produced as refinery intermediates by thermal or catalytic cracking processes can contain high levels of olefins. Olefins tend to lead to high fouling rates in pyrolysis crackers and are usually avoided as petrochemical feedstock.

There is an extensive international trade in naphtha. The above discussion illustrates that naphtha attractive to refiners may be less attractive to petrochemical operators.

Often naphtha is split at about 100°C into a heavy and a light fraction. The light fractions tend to have a higher paraffin content and more attractive to petrochemical operators and the heavy fraction containing higher levels of naphthenes and aromatics are of interest to refiners for reforming into high octane blend stock.

As well as naphtha, some operations use gas-oil as the feedstock. Gas oil is the crude oil fraction boiling typically at 220°C to 360°C, and some processing vacuum gas oils boiling typically at 360°C to 550°C. However, in some instances these crackers have been revamped to use the atmospheric column bottoms (sometimes called long residua) where the crude oil being processed has the appropriate properties of high wax (linear paraffin) content and low metal content (which otherwise promotes excessive coke formation). This material is often referred to as Low Sulphur Waxy Residual Fuel Oil (LSWR).

The typical single pass yields to the major products of interest for these feeds are shown in Table 2.3:

Table 2.3: Cracking Yields from Liquid Feedstock

	LIGHT NAPHTHA	FULL RANGE NAPHTHA	GAS OIL	VACUUM GAS OIL
boiling range (C)	36-110°C	40-164°C	176-343°C	335-515°C
methane	17.4	13.8	11.6	8.9
ethylene	**31.0**	**25.5**	**24.1**	**18.9**
propylene	**18.8**	**15.3**	**14.3**	**13.9**
C_4	10.0	8.3	8.4	9.7
py-gasoline	14.4	26.9	18.1	19.0
py-fuel oil	2.0	5.1	18.9	24.4
TOTAL	93.6	94.9	95.4	94.8
BTX content	2.6	12.1	24.1	48

The following general remarks can be made about the cracking of liquids:

- Light naphtha can produce over 30% ethylene with about half this yield of propylene. Methane yield is also high at over 17% with production of pyrolysis gasoline lower than the heavier liquids in the region of 14%. This is considerably more than the yields of pyrolysis gasoline (C_{5+} aliphatic molecules plus BTX) from gaseous feed stocks discussed above.
- Full range naphtha produces less ethylene but relatively more propylene. There is a high yield of pyrolysis gasoline.
- Gas oil produces similar yields of ethylene and propylene to full range naphtha but there is a large increase in the production of pyrolysis fuel oil (b.p. >200°C).
- Vacuum gas oil produces less olefins but relatively more propylene. The major products are pyrolysis gasoline and pyrolysis fuel oil.

An important parameter for petrochemical operations is the relative amount of ethylene and propylene in the product slate. Figure 2.5 summarises the typical relative yields of ethylene to propylene (E/P ratio) for both gaseous and liquid fuels. Also, it indicates clearly that as the feed stock gets heavier, the relative amount of propylene rises as witnessed by a fall in the ethylene/propylene ratio.

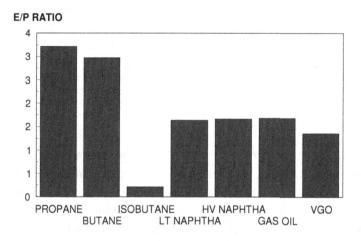

Figure 2.5: Typical ethylene/propylene ratios for various feedstocks

Reaction Severity

For liquid feedstock the product slate can be altered by changing the reaction conditions, particularly the temperature of pyrolysis or the residence time in the cracking furnace or a combination of the two. This is generally referred to as changing the severity of the cracking operation; increasing the cracking furnace temperature leading to higher severity. The higher temperature leads to more cracking and a higher yield of ethylene.

The effect on full-range naphtha is illustrated by the data in Table 2.4. This shows that increasing severity increases the ethylene (and methane) yield at the expense of propylene and heavier products.

Table 2.4: Impact of Cracking Severity on Yields

SEVERITY	LOW	HIGH
Hydrogen	2.2	3.2
Methane	10.3	15.0
Ethylene	25.8	31.3
Propylene	16.0	12.1
Butadiene	4.5	4.2
C_4	7.9	2.8
py-gasoline	27.0	22.0
py-fuel oil	3.0	6.0
TOTAL	96.7	96.6

Computer Modelling of Pyrolysis Cracking

In order to address many of the issues that have been discussed above there are available proprietary computer modelling programs which simulate commercial cracking operations. These allow the operator to simulate changes to furnace cracking operations (severity, temperature, steam ratio) and changes to feed stock including the relative amounts of components in the naphtha feed.

The modelling of naphtha cracking in particular is very complex and the simulation programs make assumptions about the overall cracking kinetics. These are modified by experience of operation in real crackers. There are several approaches to developing the models.

Of the various proprietary programs the SKF model is widely used and is reported to give good matches with commercial experience.

Differences between Pyrolysis and Catalytic Cracking

Catalysts speed up chemical processes; they do not change the position of thermodynamic equilibrium, so all of the above comments on the relative thermodynamic position of feed and product molecules applies to catalytic processes. Because catalysts accelerate chemical processes (by lowering activation energies) they can be conducted at considerably lower temperatures than pyrolysis processes and thus pyrolysis side reactions can be minimised.

For the most part we are concerned with acid catalysed reactions in which carbonium ions are the key intermediates. Carbonium ions are formed by the interaction of a feed molecule with an acid site on a catalyst surface. Carbonium ion chemistry is well defined and has several features relevant to the production of light olefins. These are illustrated in Figure 2.6.

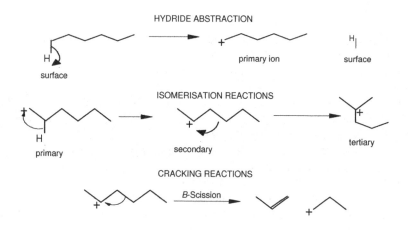

Figure 2.6: Carbonium ion reactions

The first reaction involves interaction of a hydrocarbon with the catalyst surface. Hydride abstraction occurs to form a carbonium ion. Abstraction can be of any suitable hydrogen atom but if this results in a primary ion as shown, this will rapidly isomerise by hydrogen shift to the more thermodynamically stable secondary ion. This may be further isomerised by carbon shift to a tertiary ion. This contrasts with free radicals and although isomerisation occurs it is relatively slower. The carbonium ions can also undergo inter-molecular transfer (not shown) when a carbonium ion meets another hydrocarbon molecule.

Olefins are formed from carbonium ions by β-scission reaction. This produces propylene from the secondary ion shown. Isobutene will be produced from a tertiary ion and small aliphatic molecules also produced will be branched rather than linear.

Ethylene and methane cannot be produced by β-scission and ethylene and methane are minor products that may be the consequence of some radical processes occurring within the spaces between catalyst particles.

Today much of the propylene used in the world is produced by the catalytic hydrocarbon cracking in fluid cat-cracking and similar operations[7].

---

[1] L. F. Albright, B. L. Crynes, W. H. Corcoran (eds.), "Pyrolysis: Theory and Industrial Practice", Academic Press, New York, 1983

[2] Data was adapted from D. R. Stull, E. F. Westrum, G. C. Sinke, "The Chemical Thermodynamics of Organic Compounds", Wiley, 1969

[3] D. L. Trimm in "Fundamental Aspects of the Formation and Gasification of Coke" in L. F. Albright, B. L. Crynes, W. H. Corcoran (eds.), "Pyrolysis: Theory and Industrial Practice", Academic Press, New York, 1983

[4] L. F. Albright, B. L. Crynes, W. H. Corcoran (eds.), "Pyrolysis: Theory and Industrial Practice", Academic Press, New York, 1983

[5] T. J. Ford, *Ind. Eng. Chem. Fundam.*, **25**, 240, 1986

[6] Coastal Isobutane Cracking Process developed by Foster Wheeler

[7] P. B. Venuto and E. T. Habib, "Fluid Catalytic Cracking with Zeolite Catalysts", Marcel Dekker, New York, 1979

CHAPTER 3

GASEOUS FEEDSTOCKS – PRODUCTION AND PRICE

In this chapter we discuss the methods of production, costs of production, transport and price of the gaseous feedstocks used to produce chemicals. The feedstocks of interest fall into two groups – those produced as by-products of large-scale natural gas developments for pipeline gas or LNG, and those produced from crude oil.

The first group comprise the natural gas liquids (NGLs), ethane, propane and butanes. The latter two are often referred to as a LPG and are often sold as a mixture. These feedstocks are of major interest as primary feedstock for petrochemical operations for cracking into ethylene and propylene. Liquids produced from natural gas processing are often referred to as condensate or natural gasoline. Such liquids are used in both petrochemical and reefing operations and their use as a feedstock is discussed in the next chapter.

The second group comprise LPG feedstocks made from crude oil. These are products of refinery and petrochemical operations processing heavier feeds such as gas oil and vacuum gas oil and residual fuel oils. These LPG streams contain materials of direct interest to petrochemical operations for further processing to other chemicals. With suitable treatment (hydrogenation) they can be used as cracker feedstock or sold to other users as an energy fuel.

Gaseous Feed Stocks from Natural Gas

Ethane, LPG and condensate are extracted in large amounts in the processing of natural gas. When gas comes to the surface it contains many components which need to be extracted before it can be used (for example carbon dioxide and hydrogen sulphide). The components are extracted in different unit operations, the choice and size of which is dependent upon the raw gas composition and the amount of component being extracted. Taken together, the different unit operations are referred to as the gas plant. There are many choices of the design of the gas plant and all gas plants are unique. For any gas and downstream application there are usually several viable technical and economic solutions. An analysis of gas plant design is beyond the scope of this book and we only consider the principal issues as they concern the production of NGLs. For further reference the reader is referred to the author's book[1] on "Gas Usage and Value" and Newman[2], who details 28 approaches to gas plant design, in the "Gas Processes" editions of *Hydrocarbon Processing*[3] and in the *Oil & Gas Journal*, which regularly publishes articles on gas plant design[4]. Natural gas which contains large amounts of nitrogen[5] or oxygen[6] complicates the design and increases processing costs.

Liquid products produced from gas come under a variety of names. Natural gas liquid (NGL) is a generic term for all condensed products. The C_{5+} fraction (boiling $> 30^{\circ}C$) is often referred to as condensate, or sometimes, especially in the US, natural gasoline. The C_3 and C_4 fraction is liquefied petroleum gas (LPG).

When considering the higher hydrocarbons present in natural gas it is probably best to recall that a continuum exists in oil and gas reservoirs from almost pure methane to heavy petroleum oils and waxes. In general, hydrocarbon deposits do not span to the extremes; higher hydrocarbon free natural gas is not commonly accounted although there are some very large natural gas deposits which comprise almost entirely methane[7]. Thus many oil reserves have considerable quantities of associated gas and most gas reserves have associated with them light oil (condensate) deposits.

Removal of NGLs is performed sequentially with the highest boiling fractions being removed first.

Removal of Condensate

When raw natural gas comes to the surface it is often saturated with heavier liquids. Should the gas be cooled, in an undersea-pipeline for instance, then some of the heavier components condense to form a slug of liquid in the pipeline. These liquids are removed in a series of large pipes known as a "slug catcher" – Figure 3.1.

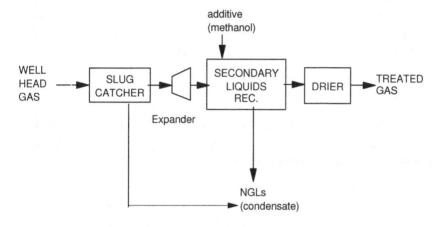

Figure 3.1: Primary gas treatment to remove condensate

Often the gas pressure is sufficient to use a turbo-expander which cools the gas to below $0°C$ and causes further condensation of hydrocarbon liquids. Because the gas stream also contains water, gas hydrate and ice formation can be a problem. This is prevented by the addition of an additive such as methanol. Following the removal of condensate, the gas stream is dried and if necessary treated further to remove acid gases such as hydrogen sulphide and carbon dioxide.

Removing excess water from the raw gas is often performed by a glycol absorption column. In some instances this facility is placed at the wellhead so that ice and hydrate formation in undersea pipelines is avoided. In a glycol dehydrator, the glycol absorbs the water which is then passed to a boiler which boils-off the water and returns cooled glycol (after heat exchange) to the absorber. There are several variants[8].

In order to remove LPG (propane and butane) from the gas stream, current processes require the gas stream to be chilled to -20°C or below. This requires the complete removal of water and carbon dioxide from the gas stream.

For the removal of the acid gases, carbon dioxide and hydrogen sulphide, many choices are available and are very dependent on the specifics of the gas and the location. Many of the options are described by Neumann[9] and Shaw[10].

After the removal of water and acid gases, the natural gas liquids can be removed. There are two main processes for the removal of LPG and ethane: turbo-expansion and refrigerated solvent absorption.

LPG Removal by Turbo-Expansion

The basic flow for a turbo-expander scheme is illustrated in Figure 3.2. This represents the simplest flow diagram, which can be quite complex if ethane is to be extracted[11].

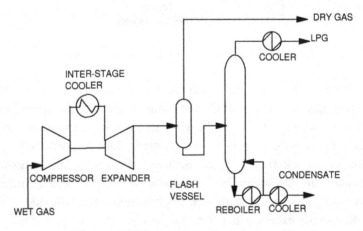

Figure 3.2: LPG extraction using turbo-expansion

Gas (known as wet gas because it contains hydrocarbon liquids to be extracted) enters the turbo-compressor and is compressed to typically 100 bar. The heat of compression is removed in an inter-stage cooler. The gas is then passed to an expander, which is coupled to the compressor in order to recover some of the required shaft power. This causes the gas stream to cool to below the liquefaction point of the LPG.

Condensed LPG and gas is passed to a flash vessel which separates the dry gas (i.e. stripped of hydrocarbons) from the liquids. The liquids are passed to a distillation column where LPG and condensate are separated. Propane and butane can be separated in an additional column and a further column is used to separate isobutane and normal butane if this is required.

Over time turbo-expander systems have improved in efficiency and can be used to extract ethane by inclusion of gas-to-gas heat recovery systems[12]. These are variously described as cryogenic systems or cold boxes and are similar in operation to the cryogenic units used for the production of LNG. The use of cold-boxes permits pre-cooling of the gas before the turbo-expander and hence an overall colder operation, this is illustrated in Figure 3.3.

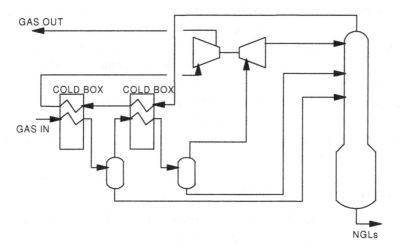

Figure 3.3: Ethane and LPG extraction using cold boxes

Inlet gas enters the first cold box. Here the gas is chilled and separated liquids are passed to a large de-methaniser column. A second cold box repeats the process after which the cold gas is expanded to condense the remaining liquids. The cold gas is now passed to the top of the de-methaniser, where it is used to cool the incoming gas in the cold-boxes.

Such systems can recover up to about 80% of the ethane present. Addition of further cooling to the top of the de-methaniser can achieve over 90% ethane recovery[13].

One point of note is that the use of cold-box technology requires the removal of mercury from the gas streams. A fire at Santos' Moomba facility in Australia in early 2004 was thought to be due to a mercury attack on the equipment

Straddle Plants

One advantage of the turbo-expander method for separating LPG from natural gas is that it allows the use of gas-pipelines to transport the LPG. LPG is costly to store and transport as it requires pressurised or cryogenic-vessels. By using gas pipelines, the lower cost transport economics of pipeline gas can be used.

In the straddle plant option, LPG is left in the sales gas at the gas-plant. The much larger volume of methane dilutes the LPG and the gas including the LPG meets the pipeline dewpoint specification. The mixture is then piped over several hundred kilometres to the straddle plant. This uses a turbo-expander to separate LPG from the gas, maintaining the residual gas within the heating value specification. There are several such operations in Canada and Australia which have been described by Hawkins[14].

Refrigerated Absorption Plants

Before the advent of turbo-expander plants in the early 1970s, the preferred method for removal of LPG materials from the gas stream was by absorption in a suitable solvent. To increase the absorption efficiencies, especially for the recovery of ethane, this technology was developed by applying refrigerated solvent to the gas stream.

The absorption plants use a hydrocarbon solvent similar to kerosene in boiling range. This is chilled to about -25°C or lower and used to absorb the required components. Because of the low

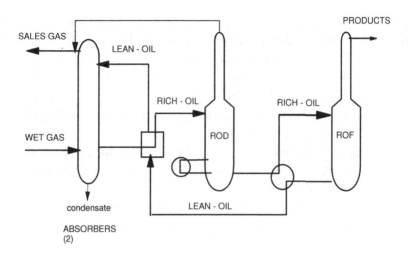

Figure 3.4: Refrigerated absorption plants – main flows

temperatures gas entering the system has to be water dry and low in carbon dioxide; these components are removed in upstream operations.

Absorption plants comprise three parts: (i) an absorber section, (ii) a section to remove dissolved gas which is returned to the sales gas stream and (iii) a distillation unit which expels the absorbed components and regenerates the solvent. Solvent free of absorbed components is referred to as lean-oil, solvent containing absorbed components is referred to as rich-oil. The main flows are illustrated in Figure 3.4.

Water dried, carbon dioxide and hydrogen sulphide free gas i.e. wet-gas, is chilled (typically to about -35°C) and enters the bottom of the absorber tower; there are usually two absorber towers. Condensate separated in the chiller unit leaves the bottom of the tower, the gas rises against a chilled falling solvent that has entered the top of the tower (the lean-oil). The solvent absorbs the heavier constituents while the lighter sales gas rises to the top of the absorber and exits the top of the tower. The now rich-oil is collected on an absorber tray above the gas entry point and passes via heat exchangers to a column (ROD).

The ROD has the duty to remove any sales gas which may have dissolved and return this to the sales gas stream. The rich-oil enters near the top of the column and falls against warmed (ca. 50 to 60°C) rich-oil circulating through a heat-exchanger. The degassed oil leaves from the bottom of the tower and the recovered gas from the top of the tower. This unit can be operated in two modes. If ethane is not a required product, the rich-oil is heated sufficiently to expel ethane (ROD means Rich Oil De-ethaniser) along with methane from the top of the tower. If ethane is to be extracted, the ROD is warmed to expel mainly methane (ROD means Rich Oil De-methaniser).

The rich oil passes to the Rich Oil Fractionator (ROF) where the solvent is boiled regenerating the lean-oil and expelling the LPG (and ethane) from the top of the tower.

As the fluids pass from the absorber to the ROF, the temperature rises from ca -35°C to about 200°C (boiling point of kerosene). This temperature difference requires extensive use of heat exchange equipment between the unit operations within the plant. Furthermore, the pressure progressively falls from about 100 atm in the absorber to about 50 atm in the ROD to less than 10 atm. in the ROF. This requires the lean-oil stream to be pumped against this pressure drop from 10 to 100 atm.

The refrigerated absorber technology is complex and manpower intensive compared to the turbo-expander technology that has largely replaced it. However, where they still exist they are particularly useful for recovering ethane, which is more difficult to extract in turbo-expander plants without refrigerated cold-boxes[15].

Case Study: Economics of Large Gas Plants

The economics of large gas plants (>1000 MMscfd gas) is of importance in understanding the production cost of ethane and LPG for petrochemical feed and to shed light on the economic drivers in refinery and petrochemicals operations. Because of the large flow of gas, these plants produce large volumes of natural gas liquids[16].

Natural gas condensate, often called natural gasoline, from these operations can be used directly as blend-stock for gasoline production. Its value to the gas plant operation is intimately linked to the prevailing price of crude oil via the value of gasoline. LPG (propane and butane) is also linked to the prevailing price of crude oil by the energy market.

There are some seasonal factors so the linkage is not as direct for LPG as is the case for gasoline (these relationships are detailed in a later chapter).

Ethane (and also propane and butane) is used as a feedstock for the production of ethylene. For this role it competes with naphtha which has a direct relationship with oil price.

Large gas plants often have the advantage that when the price of naphtha (oil) is low relative to the price of gas, ethane can be left in the gas stream and sold at the gas price thus saving the extraction cost. Conversely, in time of low gas price and high oil price, ethane can be extracted and profitably sold.

In this case study we consider a large gas plant with the following statistics, Table 3.1.

Table 3.1 Statistics for A Large Hypothetical Gas-Plant

		INPUT GAS	SALES GAS
Flow	MMscfd	1000	
	PJ/y	450	286
Methane	vol%	80.20%	94.90%
Ethane	vol%	7.00%	1.66%
Propane	vol%	4.40%	0.26%
Butane	vol%	2.30%	0.03%
C5+	vol%	3.40%	0%
Inerts	vol%	2.70%	3.19%
LIQUIDS	(t/y)	PJ/y	
Ethane	650103.9	33.72	
Propane	711569.1	35.83	
Butane	510951.7	25.3	
C5+	947071.1	46.42	

Analysis of recent published data for the construction cost of large Greenfield gas-plants indicates a cost (2007) of $1078 million[17]. The plant would have the economic parameters given in Table 3.2.

Table 3.2: Economic Statistics for a Large Gas Plant

		MMcf/d	$/GJ
WELL HEAD GAS COSTS		1000	**6.37**
	kt/y	**PJ/y**	**MM$/y**
CAPEX			1078.26
OPEX (5% CAPEX)			53.91
RECOVERY (10%DCF, 20y, FACTOR 0.143)			154.19
INPUTS			
Process Gas	8,409	427.52	2723.28
Fuel and losses (5%)	443	22.50	143.33
TOTAL feed & fuel	8,851	450.02	2866.61
OUTPUTS			
Ethane	650	33.72	
Propane	712	35.83	
Butane	511	25.30	
Sales gas	5,589	286.25	
Gasoline (C5+)	947	46.42	566.30
TOTAL	8,409	427.52	
Thermal Efficiency (%)		95.00%	
ALL PRODUCTS			
Annual Costs	MM$/y		3074.72
Unit Production Cost	$/GJ		7.19
Ethane	$/t		**373.27**
Propane	$/t		361.76
Butane	$/t		352.41
WITH GASOLINE SALES			
Annual Costs	MM$/y		3074.72
Gasoline Credits	MM$/y		566.30
Net production Costs	MM$/y		2508.42
Unit production Cost	$/GJ		6.58
Ethane	$/t		**341.61**
Propane	$/t		331.08
Butane	$/t		322.52

The typical US well head cost in 2007 was around $6.37/GJ and this has been used as the basis for the input cost in this case study. The fixed costs are the non feed operating costs which for a relatively simple turbo-expander gas plant would be about 5% per annum of the fixed capital and the capital recovery charge which is placed at 14.3% per annum of the capital (see Appendix for derivation of this value).

The process gas is supplemented by fuel gas and an operating allowance which amounts to an addition use of about 5% of the gas stream. The feedstock costs (raw input gas) dominates the costs of production.

The plant separates the components into ethane, propane, butane and natural gasoline and a sales gas which is the principal product. Two scenarios are developed: the first is when all the processing costs are assigned to all of the products including gasoline and the second is when the natural gasoline at the prevailing market price prior to distributing the costs.

Over all of the products, the production cost is $7.19/GJ. This produces ethane at $373/t. However, if the natural gasoline is sold according to the prevailing crude oil price (assumed to be $70/bbl) then this will generate by-product credit of $556 million; this is based on valuing the gasoline as naphtha with oil at $70/barrel. The basis of this oil price as a reference (index) price is discussed in the Appendix. This approach reduces the production costs and hence the unit ethane and LPG costs. The ethane production cost is $341/t.

Figures 3.5 and 3.6 give the sensitivity of the liquids production cost to input gas price and oil price, with gasoline sold prior to the distribution of the costs. The basis of the costs is in energy terms ($/GJ). This makes the product cost for ethane, propane and butane very similar and ethane is chosen as the example.

Figure 3.5 shows the sensitivity to wellhead gas price. This illustrates that for gas plants using large gas reserves for the production of LNG which requires the availability of low cost gas (typically <$2/GJ) will produce ethane below $100/tonne.

Figure 3.6 illustrates that using an input gas price of $4/GJ, there is a marked inverse sensitivity to rising oil price as the by-product credits from the natural gasoline rise.

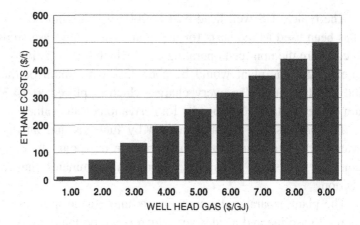

Figure 3.5: Ethane production cost and gas price; oil @ $70/bbl

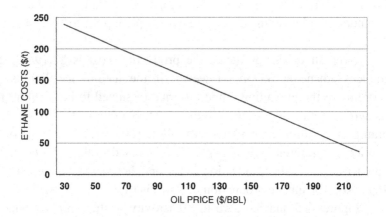

Figure 3.6: Ethane production cost and oil price; gas @ $4/GJ

The data illustrates that for an input wellhead gas price of $2/GJ or below, the production cost of the sales gas is about $2.5/GJ. Of interest are the concomitant production costs of LPG and gasoline, which are well below the prevailing prices of crude oil derived products. Selling these at prevailing oil prices makes such operations extremely profitable at high oil prices. Indeed, the combination of low input gas and high oil price produces ethane at negative costs. Plants with these statistics occur in the Middle East and other areas where there is no tangible link between the wellhead gas price and the prevailing price of energy (oil).

In these cases, such gas plants are often part of integrated refining and petrochemical complexes and offer the lowest feedstock and hence petrochemical production costs.

At a wellhead price of $6/GJ or higher, the production cost of the sales gas and the products are typically over about $6.5/GJ. This is more typical of the case in the US and Europe where the prices of wellhead gas are linked to prevailing energy prices. This lowers the operating margin of the plant.

LPG from Other Sources – LPG Quality

Most LPG is produced by gas plants. However, there are significant volumes of LPG produced by refinery operations and within petrochemical operations. These LPG streams have quite different compositions as illustrated in Table 3.3.

Table 3.3: Composition of Typical LPG Streams

COMPOUND	B.P (^{o}C)	Natural gas	Refinery (FCC)	Stream Cracker
Propane	-42.1	49%	18.00%	2%
Propylene	-47.4		17.40%	53%
Propyne	-23.2		trace	3%
Allene	-34.5			3%
cyclo-propane	-32.7			trace
n-butane	-0.5	34%	7.80%	2%
Isobutane	-11.7	17%	24.70%	3%
1-butene	-6.3		8.20%	3%
cis-2-butene	3.7		6.40%	1%
trans-2-butene	0.9		7.80%	1%
Isobutene	-6.9		9.50%	9%
1,3-butadiene	-4.4		0.30%	19%
1,2-butadiene	10.8			trace
1-butyne	8.1		trace	trace
2-butyne	27		trace	trace
but-1-ene-3-yne	5.1			trace
methyl-cyclo-propane	4 to 5			trace
cyclo-butane	12			trace

This illustrates that LPG produced from natural gas comprises only saturated molecules – propane, normal and isobutane. There is a large market for these products which are used for refinery operations, energy fuels and automotive fuel as well as for petrochemical cracking operations. The properties of these fuels, together with ethane, are given in Table 3.4.

Table 3.4: Properties of LPG Components

	ETHANE	PROPANE	n-BUTANE	iso-BUTANE
Mol Wt	30.07	44.1	58.1	58.1
BP (C)	-88.6	-42	-0.5	-11.7
MP (C)	-183.3	-187.7	-138.4	-159.6
RVP (psia)	-800	190	51.6	72.2
s.g. (liq)	0.3564	0.5077	0.5844	0.5631
HHV (gas, GJ/t)	51.9	50.4	49.5	49.4
HHV (BTU/cf)	1768.8	2517.5	3262.1	3252.7
RON	111.5	112	93.8	101.5
MON	101	97.1	89.6	97.6

Refinery LPG, typically produced from a fluid cat-cracker unit, contains in addition to paraffins a large amount of olefins particularly propylene and isobutene. These olefins and isobutane have refinery uses and are often used in producing additional fuel such as polymer gasoline and alkylate. Also present are trace quantities of dienes and acetylenes.

If LPG containing olefins and diolefins is to be used as a feed for a cracking operation then it should be hydro-treated prior to use. This will prevent olefin polymerisation in the cracking furnace which would lead to coking. Hydro-treatment of LPG is becoming more common in refinery operations as the specifications for automotive LPG are tightened[18].

Changes to refinery operations in countries requiring the production of high quality gasoline has altered the balance of LPG in many refineries and many produce LPG for vehicle use or petrochemical use. Potentially LPG from refineries can be contaminated with dienes which can lead to excessive coke lay-down in cracking operations.

The LPG stream produced by pyrolysis cracking contains all possible C_3 and C_4 molecules. As well as olefins, prominent are highly unsaturated materials such as acetylenes and dienes.

Sometimes propane from natural gas and refinery operations becomes contaminated with carbonyl sulphide (COS) which is not removed in acid gas plants. Alternative approaches to removing COS from liquid propane by sorption processes have been compared by Wilson *et al.*[19]

Ethane and LPG (Propane) from LNG

The enormous growth in the world trade of LNG (liquefied natural gas) is leading to the idea that re-gasification could generate significant volumes of ethane and propane. Many of the world's LNG operations leave significant amounts of ethane and propane in the LNG in order to meet the heating specification demanded by the many of the world's LNG importers, particularly those in the Far East. However, some jurisdictions, in particular the US and Europe, in respect to the gas specification requires the ethane and propane stripped from the gas prior to distribution. Since the heavier components, ethane and propane, can constitute 10% of the mass of the LNG then significant volumes of feed could become available by this route[20].

Use of LPG in the Chemical Industry

The various components of LPG streams are used in a variety of processes. Propane, butane and isobutane are used as cracker feedstock for the production of olefins which is discussed in later chapters. In addition n-butane is used for the production of 1,3-butadiene. This compound can also be extracted from the C_4 cracked gases by extensive distillation coupled with a selective absorption process.

Butene is used as a co-monomer in the production of LLDPE and the production of some speciality polymers and ethers. It can be extracted from C_4 cracked gases by distillation or by dehydrogenation of butane.

Isobutene is also used to produce MTBE. Often the entire C_4 cracked gas or FCC C_4 stream, which contains isobutene, can be used as a feed for the MTBE plant without the need to extract the isobutene.

Prices of Gaseous Feed Stocks

Condensate, or natural gasoline, is directly linked in value to that of crude oil. In many parts of the world it sells at a discount to the local marker crude oils because its boiling range profile does not easily fit into normal refinery operations – it contains too much light boiling fractions relative to typical crude oils. The discount is typically about $1/bbl but for a given condensate there is considerable variability in the differential. If the octane is sufficient it may be used as a gasoline blend stock and this may make it more attractive than crude oil in some circumstances.

Condensate is often passed to a splitter column and distilled into light and heavy naphtha for petrochemical operations. This is discussed further in a later chapter.

Ethane prices are generally determined by local circumstances. The floor price for ethane is set often priced according to the price of gas on an energy basis. For example in the US for flexible fuel cracking operations can use both ethane and naphtha; if demand falls then ethane can be left in the gas stream and sold as gas. The US Energy Information Administration collates data for the well-head gas price. The data is shown in Figure 3.7.

This graph shows collated data from across the US. Prior to about 2000, apart from a few spikes, the prices ranged typically in the range $1 to $2/Mcf. Since 2000, gas prices have been very volatile and as with the rise in energy prices over the period 2003 onwards there has been a dramatic rise in the price of well head gas with some extreme peaks over $10/Mcf.

The ethane "market" price is set by demand, which is influenced by the relative ratio of oil to gas. In large markets with flexible fuel cracking operations (US, EU), if oil price is high, ethylene producers switch to ethane feedstock. If oil price is low, ethylene producers switch to oil (naphtha). However, excess by-products from naphtha can put a limit onto the extent of the switch.

For some countries the cracking operation is based entirely on ethane and petrochemical operators enter take or pay contracts for ethane. Often there is a fixed-variable component in the contract linking ethane price to the prevailing price of crude oil. Obviously this limits the benefits to the operator in times of rising oil price with some or all of the benefit passed on to the ethane supplier.

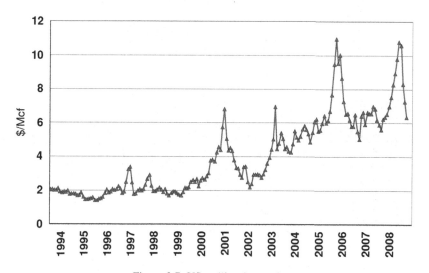

Figure 3.7: US wellhead gas price

Propane and butane (LPG) price set by reference to the prevailing LPG market. There is a very large trade in LPG in the major economies. Its main use is as a commercial energy fuel it is used in very large amounts to this end. Most of the big producers in the US or the North Sea sell to the local markets. This leaves Saudi Arabia as the major swing producer which sells according the supply and demand across the world. The consequence is that most LPG prices are set relative to the Saudi Aramco contract price which is set on a monthly basis. The history of the propane Aramco and US propane price is illustrated in Figure 3.8.

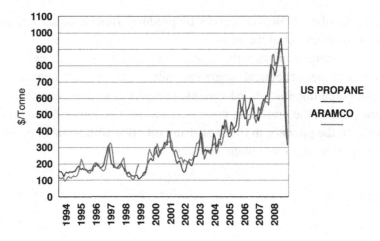

Figure 3.8: Aramco and US propane prices

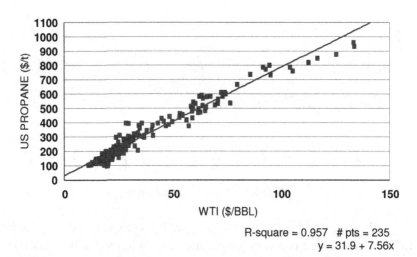

R-square = 0.957   # pts = 235
y = 31.9 + 7.56x

Figure 3.9: US propane correlation with oil price (WTI)

The graph illustrates that Aramco and US propane prices are generally in step with the price ratio average about unity. The graph also indicates the general rise of prices with time. Since much of the world's LPG is used for heating purposes, there is a reasonable correlation with the prevailing local crude oil marker price. This is illustrated in Figure 3.9 which plots the US propane price against WTI.

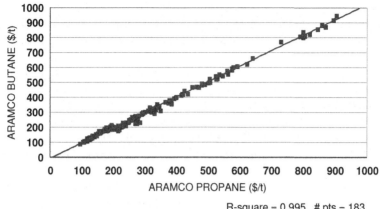

Figure 3.10: Aramco butane and propane correlation

Butane is priced similarly to propane and there is a strong correlation between propane and butane prices as illustrated in Figure 3.10 for the Saudi Aramco Contract.

[1] D. Seddon, "Gas Usage and Value", PennWell, Tulsa, Oklahoma, 2006

[2] S. A. Newman (ed.) "Acid and Sour Gas Treating Processes", Gulf Publishing, Houston, Texas, 1985

[3] Gas Processes is published biennially in even years in the April/May editions of *Hydrocarbon Processing*

[4] For example A. Habibullah, *Oil & Gas Journal*, June 3, 2002

[5] M. J. Healy, A. J. Finn, L. Halford, *Oil & Gas Journal*, February 1, 1999, p. 36; M. Mitariten, *ibid.*, April 23, 2001, p. 42; A.S. Padron, J. E. R. Rodriguez, E. B. Vazquez, A. J. A. Senosiain, G. E. M. Tapia, *ibid.*, December 1 2003, p. 50

[6] M. Howard, A. Sargent, *Oil & Gas Journal*, June 23, 2001, p. 52

[7] Coal bed methane (coal seam gas) contains very little hydrocarbon higher than methane.

[8] D. W. Choi, *Hydrocarbon Processing*, July 2006, p. 69

[9] S. A. Newman (ed.) "Acid and Sour Gas Treating Processes", Gulf Publishing, Houston, Texas, 1995

[10] T. P. Shaw and P. W. Hughes, *Hydrocarbon Processing*, May 2001, p. 53

[11] V. Aggarwal, S. Singh, *Hydrocarbon Processing*, May 2001, p. 41; Y. R. Mehra, *Oil & Gas Journal*, October 29, 2001, p. 56; K. L. Jibril, A. I. Al-Humaizi, A. A. Idriss, A. A. Ibrahim, *ibid.*, March 6, 2006, p. 58

[12] R. Chebbi, A. S. Al-Qaydi, A. O. Al-Amery, N. S. Al-Zaabi, H. A. Al-Mansouri, *Oil & Gas Journal* , January 26, 2004, p. 64; R. Chebbi, K. A. Al Mazroui, N. M. A. Jabbar, *ibid.*, December 8, 2008, p. 50

[13] A. A. Rahman, A. A. Yusof, J. D. Wilkinson, L. D. Tyler, *Oil & Gas Journal*, October 25, 2004, p. 58; R. Chebbi, A. S. Al-Qaydi, A. O. Al-Amery, N. S. Al-Zaabi, H. A. Al-Mansouri, *ibid.*, January 26, 2004, p. 64

[14] D. J. Hawkins, *Oil & Gas Journal*, December 16, 2002, p. 46; *ibid.*, December 23, 2002, p. 54; *ibid.*, January 6, 2003, p. 48; *ibid.*, January 20, 2003

[15] S. M. Al-Shahrani and Y. M. Mehra describe an absorbent system for LPG extraction in *Oil & Gas Journal*, June 4, 2007, p. 60

[16] K. L. Currence, B. C. Price, W. B. Coons, *Oil & Gas Journal*, April 12, 1999, p. 49

[17] Authors analysis of *Hydrocarbon Processing* "Boxscore" data

[18] S. Habibi, J. Nava, *Hydrocarbon Processing*, July 2007, p. 75

[19] S. Wilson, R. Kimmitt, R. B. Rhinesmith, *Oil & Gas Journal*, September 22, 2003 based on presentation to 82nd GPA Convention, March 9–12, 2003, San Antonio

[20] S. Huang, D. Coyle, J. Cho and C. Durr, *Hydrocarbon Processing*, July 2004, p. 57; K. Otto, *Hydrocarbon Asia*, March/April 2005, p. 20

CHAPTER 4

LIQUID FEEDSTOCK, PRODUCTION AND PRICE

This chapter considers the production and price of liquid feedstock of interest to the petrochemicals industry. This mainly concerns naphtha, gas oils and residual fuel oils for both feedstock and energy. These feedstocks are produced by the primary operations in oil-refining. As a consequence many petrochemical complexes are juxtaposed to refineries. For other operations there is a large trade in the required materials and the feedstock can be purchased on the open market.

Primary Refinery Operations

Here we consider the refinery production of petrochemical feedstocks. Downstream refinery processes will only be discussed as it applies to the quality of these feeds.

The primary processes of a refinery operation are illustrated in Figure 4.1.

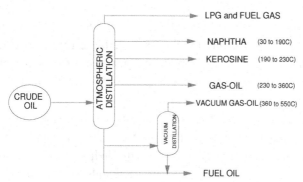

Figure 4.1: Primary products from crude oil refining

Crude oil enters the refinery and is stored and excess salt removed prior to distillation. Often several different crude oils are co-mingled before distillation in the atmospheric distillation tower. This has the duty to separate the crude oil into fractions according to distillation range.

LPG leaves the top of the tower and is passed to a gas-plant where it is mixed with similar streams from other refinery processes. The naphtha fraction comes next, boiling between about 30°C and 190°C. Often this stream is split into a light-naphtha (boiling to about 30°C to 100°C, often called straight-run gasoline) and a heavy naphtha stream. In refineries the light naphtha is blended with other streams to produce gasoline and the heavy naphtha stream is reformed to produce a high octane stream.

Boiling higher than naphtha is the kerosene fraction, boiling typically 190 °C to 230°C. This fraction is used for the production of jet-fuel.

The next boiling fractions are the gas-oils, which in the refinery context are used to produce diesel. In the atmospheric column, the boiling point of the heaviest fraction is about 360°C.

The column bottoms are known variously as atmospheric residual fuel oil or long residua and they contain all of the materials boiling higher than about 360°C, including contaminant metals. This material is often sold as a light fuel oil. If the metal content is low and there is a high wax content, it can be used as a petrochemical cracker feedstock in an appropriately configured steam cracking operation.

Shown in Figure 4.2 is a simple refinery flow-sheet. There are many refineries in the world configured in this manner. These are often referred to as "simple" refineries. In many of these refineries, the crude oil column has greater capacity to the downstream processing units and the refinery sells the excess intermediate streams such as naphtha on the oil market.

A point of note is that refiners, in meeting the various fuel specifications, have some flexibility in the distillation cut points and often change these on a regular basis. This helps the refiner to better match the output of the column to the demands of the transport fuel market. For example, if the refiner is faced with an increase in demand for jet fuel, then he can increase the volume of jet production by lowering the top cut point for naphtha (to 180°C say rather than the usual 190°C) and increase the top cut point (from 230°C to 240°C say). There may be concomitant changes in the distillation profile of intermediate streams exported from the refinery.

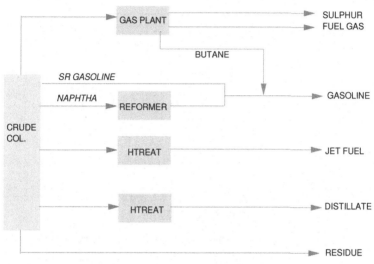

Figure 4.2: Simple refinery flow-sheet

In many refineries the atmospheric column bottoms are passed to vacuum distillation. This produces vacuum gas oils used primarily to produce lubricating oils. These boil in the range 350°C to about 550°C. The residua from this column, often referred to as short residua or heavy fuel oil, concentrates all of the contaminant metals.

Sulphur has to be removed for many petrochemical operations. All crude oils contain sulphur. The sulphur is distributed throughout the boiling range. The amount and type of sulphur present depends on the source crude oil. This is illustrated in Figure 4.3 for Saudi Arabian light and heavy crude oils which contain 1.77 and 2.8 wt. % sulphur respectively.

This plot shows that most of the sulphur is concentrated in the heavier fractions. Even for crude oils with high sulphur content the level of sulphur in the naphtha fraction can be low (below 100 ppm). This makes the naphtha (commonly referred to as straight-run naphtha) a particularly attractive feedstock and is used widely to produce synthesis gas (a mixture of carbon monoxide and hydrogen) for the production of ammonia, methanol and oxo-alcohols when other low sulphur feed stock such as natural gas is unavailable[1].

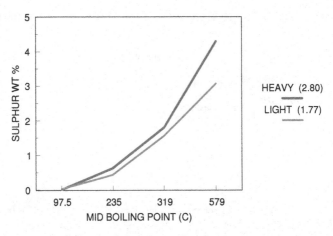

Figure 4.3: Sulphur content and crude fraction; Saudi Arabian crudes

"Complex" refineries build on the simple refinery flow-sheet with refinery operations with the duty to convert residual fuel into lighter products (naphtha, kerosene and diesel). Many of these downstream processes as well as cracking the heavier compounds also introduce sulphur in the lighter streams. Thus naphtha produced from coking operations and fluid-cat-crackers are generally higher in sulphur than the straight-run naphtha. Depending on the fuel quality being produced this may not be a problem for the refiners but sulphur contamination can be a problem for petrochemical operations.

Condensate Splitters

Naphtha is also produced from natural gas condensate by distillation[2]. In many condensates there exists a long high boiling tail which makes it unsuitable for pyrolysis cracking. Further, the relatively lower value (in mass terms) of condensate relative to naphtha makes it economically attractive to distil the condensate. In a typical condensate splitter, shown figuratively at Figure 4.4, condensate is distilled into a light and heavy fraction. Light materials and high boiling column bottoms are used to fuel the distillation furnace if not otherwise used in other processing activities.

The economics of condensate splitters are variable and depend to a large extent on an attractive differential between the condensate and naphtha products.

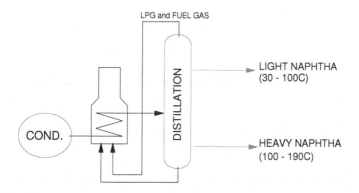

Figure 4.4: Condensate splitter

Naphtha Quality

The quality of naphtha reflects its source crude oil or condensate or the refining process that has produced it. From the overview given in Chapter 2 concerning the selection of naphtha for cracking, it is seen that petrochemical cracking operations are best served by naphtha with high linear paraffin and low aromatic content, whereas refinery operations are best served with naphtha of high branched paraffin and high aromatic content. Table 4.1 collates the typical properties of various types of naphtha produced from the distillation of crude oil.

In order to assist buyers of naphtha, parcels are characterised in terms of this PNA (paraffin, naphthenes and aromatics) analysis. In addition some organisations use the UOP or Watson K-factor as a characteristic.

Table 4.1: Typical Naphtha Properties and Japanese Open Spec.

FRACTION	BP (C)	DENSITY (kg/l)	RON	P	N	A	K-factor
FULL RANGE	C5 to 200	0.725	52	57	30	13	12.1
LIGHT NAPHTHA	C5 to 100	0.668	70	82	15	3	12.6
HEAVY NAPHTHA	100 to 200	0.754	52	55	32	13	11.9
HEAVY 1	100 to 150	0.734	61	55	35	10	12
HEAVY 2	150 to 200	0.773	43	49	36	15	11.8
JAP SPEC	24 to 204	0.665 to 0.740		65 MIN			

The UOP K-factor attempts to judge the "paraffinicity" of a fraction. It can be used for any petroleum fraction as well as naphtha. It is defined as:

$$K = T_b^{1/3} / s$$

Where T_b is the molal average boiling point of the fraction in degrees Rankin and s is the specific gravity of the fraction. The K-factor can be correlated with other physical parameters of the fraction: API gravity and viscosity; API gravity and flash point; API gravity and aniline point; flash point and refractive index.

Paraffinic fractions have K-factors of about 12.5; naphthenes have K-factors of about 11.5, whereas aromatic fractions have K-factors of about 10.

The Japanese OPEN Specification

All of Japan and Korea's large petrochemical industry has been built around the cracking of naphtha. This has generated a large trade in naphtha for petrochemical cracking. The naphtha traded conforms to the so called Japan "open" specification; given in Table 4.1. As may be deduced, the achievement of this specification is not unduly arduous by the majority of straight-run naphthas.

Refinery Intermediate Naphthas

Occasionally naphtha originating from downstream refinery operations comes onto the naphtha market. The properties of some of these naphthas (Thermal, Vis-breaker, Coker and Fluid cat-Cracker) are given in Table 4.2. These intermediate naphthas contain high olefin content and higher sulphur content than straight-run and should be avoided in cracking operations because of coke lay-down in the cracking furnace. Hydro-cracker naphtha has zero sulphur and no olefins but can still contain significant amounts of naphthenes and aromatics.

Table 4.2: Typical Naphtha Properties (wt%)

	S%	P	O	N	A	RON
Straight run	0.01	50	1	30	19	55
Thermal	0.6	45	25	15	15	75
Vis-breaker	3	23	45	11	11	
Coker	2	15	55	3	27	78
Cat-Cracker	0.2	33	44	2	21	91
Hydro-cracker	0	17	0	41	42	84
FT Naphtha SR	0	> 85	< 10	< 2.5	< 2.5	30
FT Naphtha HT	0	> 95	0	< 2.5	< 2.5	20

The properties in the table are representative and there is a range for each type. Straight-run usually contains 100ppm sulphur or less but there are some exceptions. Generally straight run has a high level of paraffins (P), few if any olefins (O) and a varying amount of naphthenes (N) and aromatics (A). The octane rating (RON) is typically 55 or higher. Some straight-run naphtha contains high levels of aromatics which do not make good cracker feed. It has been proposed that such naphtha could be pre-treated to remove the aromatics prior to the cracking operation. This would improve ethylene yields and provide additional aromatics for downstream operations[3].

In recent years there is an interest in converting natural gas or coal into high quality diesel fuel by the Fischer-Tropsch process (FT). This produces a significant by-product yield of naphtha with high paraffins and a poor octane. The straight-run can contain some olefins (typically 10%) and some oxygenates. With post production hydrogenation, which is usually the case, these are eliminated and the hydro-treated products (FT naphtha HT) contain almost entirely paraffins making excellent cracker feedstock.

Mercury, Sulphur and Other Contaminants

Because many petrochemical operations use cryogenic separation to separate hydrogen in the cracked streams, it is important to maintain the stream free of mercury. Mercury can contaminate naphtha, especially if it is derived from natural gas condensate since traces of mercury can be found in most natural gas[4]. Mercury in naphtha is readily removed using carbon sieve technology[5].

For some uses, even traces of sulphur is a problem and for reforming there is a wide range of trace contaminants which poison the precious metal reforming catalyst. For reforming and some cracker operations, naphtha is hydro-treated immediately prior to use in order to reduce the level of contaminant to an acceptable level or to assure the durability of a downstream operation.

Since the addition of a distillation column after hydro-treatment is relatively easy, sometimes natural gas condensates can be used as a primary feed to cracking operations.

Price of Naphtha and Other Liquid Feed Stocks

Most naphtha (over 80%) is used in the production of gasoline. Therefore the price of naphtha is strongly influenced by, on the one hand, the prevailing price of crude oil and on the other, by the demands of the gasoline market. The trade in oil, naphtha and gasoline is very large and transparent.

Crude Oil Prices

Broadly, there are three major world market centres – New York for the Americas, Rotterdam for Europe and Singapore for the Far East.

Each market has its local "marker" crude to which other crude oils are referenced – West Texas Intermediate (WTI), Brent Blend and Tapis Blend respectively. There is extensive trade between the regions (arbitrage) which links the global oil and oil derivatives market into one structure and evens out regional price differentials.

These representative crude oils are of the group of light low sulphur crude oils, which are easily processed into high quality transport fuels. They are sought for these properties and sell at a positive differential to most other crude oils.

Recent experience in the oil price is illustrated in Figure 4.5 for three marker light and low sulphur crude oils – WTI, Brent and Tapis considered the reference crude oils in the three regional markets. The graph shows that the three crude move in unison to events in the oil market[6].

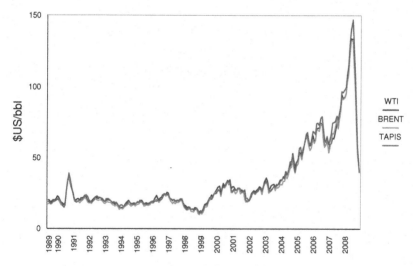

Figure 4.5: Crude oil prices

Late 1990 saw the first Gulf War which caused a short term spike in the price of crude which rose to nearly $40/bbl. This was followed by a progressive decline in its price to a low point of about $10/bbl in late 1998. Since that time there has been a progressive rise in the price of oil which has accelerated since 2003 to reach over $60/bbl in 2005 then a period of very high price volatility to reach highs of $150/bbl in mid 2008 before a collapse in the last quarter of the year.

The close correlation between the marker crude oil prices is illustrated in Figure 4.6 and Figure 4.7 where Brent and WTI are correlated to Tapis crude.

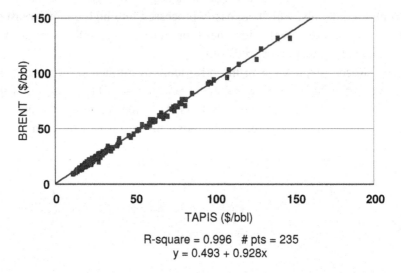

R-square = 0.996   # pts = 235
y = 0.493 + 0.928x

Figure 4.6: Correlation of Brent and Tapis

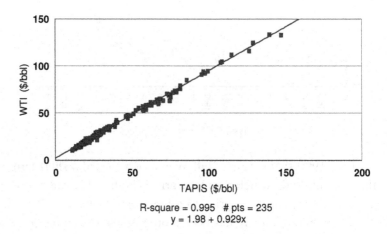

R-square = 0.995   # pts = 235
y = 1.98 + 0.929x

Figure 4.7: Correlation of WTI and Tapis

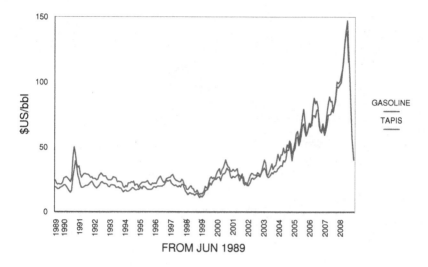

Figure 4.8: Gasoline and Tapis (Singapore)

Careful examination of the oil price history (Figure 4.5) sees that during the early part of each year, strong demand in the US (WTI) causes a small peak in the price of oil which feeds into the other oil markets. This is due to refiners producing stocks of gasoline for the US driving season in July and August. This is further illustrated in Figure 4.8 which shows the gasoline and Tapis prices on the Singapore market showing a price hike early in the year.

Naphtha Prices

The price of naphtha is strongly linked to the price of crude oil and will be influenced by the gasoline market. This is illustrated in Figure 4.9 where the European price of naphtha reported by *ECN News* is correlated with the price of Brent crude oil price reported by the US EIA. Note the excellent correlation with a correlation coefficient (R^2) of about 0.98.

Gas Oil and Residual Fuel Oil

To illustrate the price issues concerning other petrochemical feed stocks, the monthly average price data for the Singapore market is

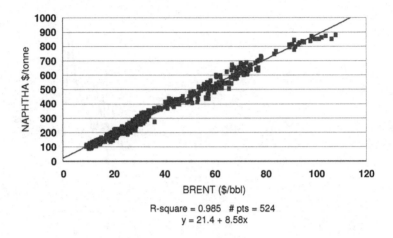

R-square = 0.985   # pts = 524
y = 21.4 + 8.58x

Figure 4.9: Correlation of naphtha and Brent

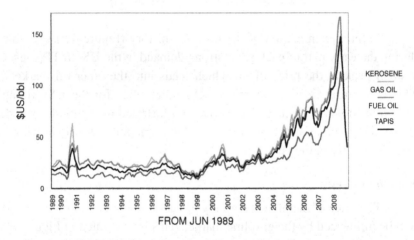

Figure 4.10: Kerosene, gas oil, fuel oil and Tapis (Singapore)

discussed. The historical price trends for kerosene, gas-oil, fuel oil (180 cSt) and Tapis are illustrated in Figure 4.10.

The figure illustrates that all the products fall and rise in unison. However there is some discrepancy with fuel oil. The correlations between gasoline, kerosene and gas oil, fuel oil and Tapis are illustrated in Figures 4.11, 4.12, 4.13 and 4.14.

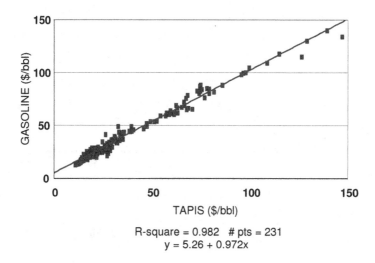

Figure 4.11: Correlation of gasoline and Tapis (Singapore)

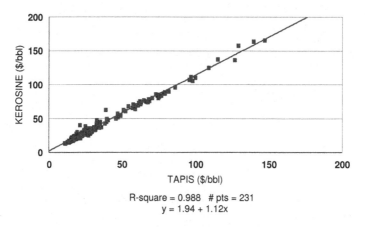

Figure 4.12: Correlation of kerosene and Tapis (Singapore)

Some crude oils are low in sulphur and waxy. Crude oil of this nature is common in South Asia and is processed in large "simple" refineries. They produce a fuel oil which is referred to as low sulphur residual fuel oil (LSWR) for export. LSWR generally sells at a premium to fuel oil, but the price differential is seasonal with a major market being

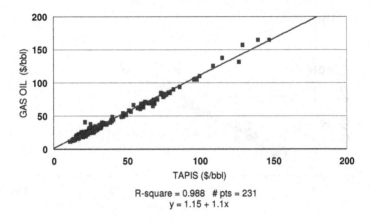

Figure 4.13: Correlation of gas oil and Tapis (Singapore)

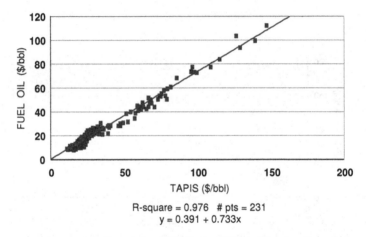

Figure 4.14: Correlation of fuel oil and Tapis (Singapore)

power generation in countries with strict limits to sulphur emissions where it is used in place of conventional, higher sulphur fuel oil in the summer months of high electricity demand (particularly in Korea and Japan). LSWR also differs from conventional fuel oil in having a high pour point, typically 40°C. This makes it difficult to store and ship (requires heating).

LSWR is finding increasing use in cracking operations, particularly those configured to crack gas oil. The high wax content, indicative of linear paraffins, generates a good ethylene yield and the pyrolysis fuel oil is low in sulphur and used to produce carbon black.

---

[1] D. Seddon "Gas Usage and Value", PennWell, Tulsa, Oklahoma, 2006

[2] FACTs Inc report on production and condensate splitters is reviewed in *Oil & Gas Journal*, Feb. 20, 2006, p. 50

[3] V. L. Bhirud, *Hydrocarbon Processing*, Apr. 2007, p. 69

[4] McNamara and Zavora, *Hydrocarbon Asia*, May/Jun. 1997, p. 76

[5] K. Ikushima, N. Akihisa and S. Matumoto, *Hydrocarbon Processing*, Dec. 2006, p. 97; D. Young and P.J.H. Carnell, *Hydrocarbon Asia*, Jul/Aug 1006, p. 42

[6] During the early part of 2009 specific features in the US oil market have tended to change the pattern in the relative value of WTI versus similar crude oils

CHAPTER 5

VALUE OF PRODUCTS, STORAGE AND TRANSPORT

In this chapter we consider two subjects which impact the economic viability of large integrated chemical complexes, the value of products and by-products produced and the transport of product and by-product to a distant destination.

Products

We are primarily concerned with the production of the light olefins ethylene and propylene. In many parts of the world these products can be sold directly to a user within a petrochemical complex or to third party users by pipeline. In these cases there is no or minimal transport cost to be considered.

Traded prices for ethylene and propylene produced and sold on the pipeline network in Europe are shown in Figure 5.1[1]. This plots the spot prices for ethylene and propylene from 1989 to 2008 with the price of naphtha as bars underneath. The graph shows the following features:

- Ethylene is generally at a higher price than propylene. This is not always the case and for the Far East propylene is generally worth more than ethylene.
- Relative to the price of naphtha, the price of olefins is far more volatile with large peaks and troughs.
- The peaks and troughs represent business cycles in the petrochemical business; typical peak to peak values are 1.5 to 2.5 years.

87

- There are many periods of low price with the traded price in the vicinity of $300/t. This low price can last for long periods. This effectively sets the floor price which a petrochemical operation should aim to beat - that is have a production cost below the floor price.
- From 2003 to late 2008 there was a progressive rise in the price of oil and hence naphtha. Over the period the price of olefins also rose. In late 2008 both oil price and olefin prices collapsed with olefins heading towards the floor price.

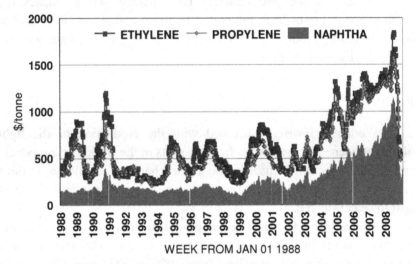

Figure 5.1: Olefin and naphtha price trends

Hydrocarbon By-Products

Successful petrochemical projects are characterised by either eliminating the production of by-products, or if they have to be produced, by maximising the value of by-products from the process. Here we are primarily concerned with maximising by-product credits.

The basic value of a hydrocarbon by-product is its value as a fuel oil substitute for heating process furnace operations. Since fuel oil is generally the hydrocarbon of lowest value, degrading by-products to fuel value will result in a cost to the process – high value feedstock is degraded to fuel oil. This should be avoided if possible, but for some by-products such as methane from cracking operations its use as a fuel oil substitute (i.e. fuel gas) may be the only option.

A higher value option for hydrocarbons may be to use them as a feedstock substitute. In cracking operations, the product slate from the primary cracking operations contain a lot of ethane, even for naphtha and gas oil cracking. Rather than degrade the ethane to fuel oil value, ethane is separated and recycled to a special cracker-furnace which cracks the ethane to ethylene. The cracker-furnace is separate to the main naphtha cracking furnaces because, as discussed in Chapter 2, the cracking of ethane requires a higher temperature than larger naphtha molecules.

A number of petrochemical processes produce significant volumes of hydrogen as a by-product, including pyrolysis cracking. This can be used as a fuel oil substitute, but this greatly undervalues hydrogen, and alternative use in other chemical processes is the better option and generally pursued by successful operations.

Following is a discussion about the by-products from the various naphtha cracker streams:

C_2 stream

The product of interest is ethylene and this is contaminated with ethane and acetylene. The most common practice is for acetylene to be selectively hydrogenated to ethylene using supported palladium catalysts:

$$C_2H_2 + H_2 = C_2H_4$$

This process produces a small amount of by-product "green oil" which is degraded to fuel oil[2].

Acetylene itself has considerable value (equivalent to ethylene) for producing a variety of specialist chemicals as well as the commodity monomer vinyl chloride (by addition of hydrogen chloride) and its use as a specialist fuel – acetylene welding. Acetylene is very dangerous in the liquid state, and is not distilled. In the pure form it has a tendency to explosively decompose:

$$C_2H_2 = 2C + H_2$$

The concentration of acetylene in the C_2 stream can be increased by increasing the severity of cracking. The acetylene is then extracted using a solvent extraction process (copper solutions are a common method) to separate the acetylene component from the C_2 stream.

C_3 Stream

The product of interest is propylene and propane which are contaminated with propyne (methyl acetylene), allene and *cyclo*-propane. A typical composition is given in Table 5.1.

Table 5.1: Typical Composition of a Naphtha Cracker C_3 Stream

COMPOUND	B.P ($^\circ$C)	Stream Cracker
Propane	-42.1	2%
Propylene	-47.4	53%
Propyne	-23.2	3%
Allene	-34.5	3%
cyclo-propane	-32.7	Trace

There are few significant industrial uses[3] for these other materials and they are reduced by selective hydrogenation to propylene and propane.

Propane is separated by distillation and can be either recycled to produce cracker feedstock or purified to a saleable LPG product.

C_4 Stream

A cracker C_4 stream contains all of the possible C_4 hydrocarbons which are listed in Table 5.2. Of these commercial interest focuses on butenes, isobutene, 1,3-butadiene and butanes. Efficient separation is impossible by distillation alone and complete separation is by a combination of distillation, selective hydrogenation and selective absorption. If butadiene is not required this can be hydrogenated and the butenes and butane separated by distillation.

Table 5.2 Typical Composition of a Naphtha Cracker C$_4$ Stream

COMPOUND	B.P (°C)	Stream Cracker
n-butane	-0.5	2%
Isobutane	-11.7	3%
1-butene	-6.3	3%
cis-2-butene	3.7	1%
trans-2-butene	0.9	1%
Isobutene	-6.9	9%
1,3-butadiene	-4.4	19%
1,2-butadiene	10.8	trace
1-butyne	8.1	trace
2-butyne	27	trace
but-1-ene-3-yne	5.1	trace
methyl-cyclo-propane	4 to 5	trace
cyclo-butane	12	trace

Both 1-butene and 2-butene can be used as a monomer for specialist polymers. Of interest to integrated cracking and polymer production operations is 1-butene for co-polymerisation with ethylene to produce LLDPE (linear low-density polyethylene)[4]. For ethane cracking operations where the C$_4$ stream maybe insufficient, 1-butene can be made by from ethylene by dimerisation[5].

Isobutene is used for the production of MTBE (methyl tertiary butyl ether) which nowadays is used little in the US market but is widely used as a gasoline octane booster in many countries.

The extraction of butadiene involves solvent extraction and distillation. In the process shown in Figure 5.2[6], a mixed C4 steam enters a solvent stripping column (1) which strips the butadiene and acetylene compounds from the stream. A typical solvent is N-methylpyrolidone (NMP).

A rectifying column (2) removes all of the butenes from the crude butadiene stream part of which is sent to a second solvent stripper (3) with the bottoms containing C$_4$ acetylene compounds returned to the rectifying column. The bottoms from the rectifying column (2) are fed to the solvent stripping column (4) which returns lean solvent.

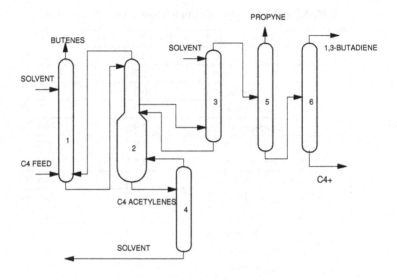

Figure 5.2: Butadiene extraction from C_4 streams

The butadiene is purified in two columns which removes propyne (methyl acetylene) and C_{4+} products which are mainly 1,2-butadiene and C_5 hydrocarbons.

1,3-Butadiene has a considerable demand for use in the production of synthetic rubber and acrylonitrile-styrene-butadiene (ABS) co-polymers. It is difficult and costly to separate from the mixed C_4 stream and results in few cracking operations building the necessary plant. The continued growth in demand[7] for butadiene and closing of several process plants has resulted in a dramatic growth in the value of butadiene in recent years; Figure 5.3. The figure dramatically illustrates that for over a decade the average butadiene price was below $500/tonne but since 2004 there was a steady rise before a dramatic rise in value in 2007/8 before the price collapses in late 2008.

The final product of interest is butane. This can be separated and either sold as LPG or recycled as a cracker feedstock. All of the C_4 stream can be recycled for cracking. However, olefins and especially dienes and the C_4-aceylenes rapidly form coke and the C_4 stream is generally fully hydrogenated to butane.

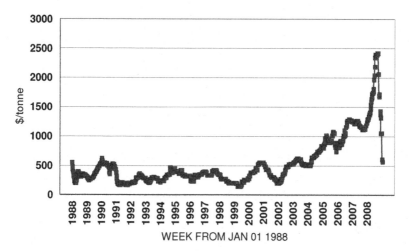

Figure 5.3: Butadiene price trends

$C_5 - 190^oC$ Stream – Pyrolysis Gasoline

The C_5 to 190°C stream is often referred to as pyrolysis gasoline. In naphtha cracking it forms the dominant portion of the liquid products. The unit value of gasoline is higher than that for naphtha so that production and sale of pyrolysis gasoline is a profitable option. In the large operations which are integrated with refineries, the pyrolysis gasoline is used as a gasoline blend-stock. It can also be used directly as gasoline in countries which do not require the production of high quality transport fuels. For other producers pyrolysis gasoline is sold on the open market usually at a discount to the prevailing gasoline price.

Pyrolysis gasoline contains a large quantity of aromatics – typically >60% benzene, toluene and mixed xylene (often referred to as BTX) – which imparts to the fuel a high octane level (typically >95 RON); Table 5.3. Unfortunately, the main component is benzene, which is no longer favoured as a gasoline component which has led to a decline in its use.

However, the aromatics, in particular benzene, are highly sought after as petrochemical intermediates and gasoline additives. Recent price trends are illustrated in Figure 5.4.

Table 5.3: Typical Composition of C_5-190°C Product from Naphtha Cracking

Benzene	40.00%
Toluene	20.00%
Xylene	7.00%
Aliphatic hydrocarbons	33.00%
TOTAL	100.00%

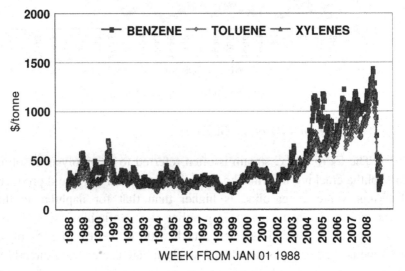

Figure 5.4: BTX price trends

The figure illustrates that benzene generally sells at a premium to toluene and xylene reflecting its use as a chemical intermediate in the production of styrene, phenol and nylon[8]. The floor value for toluene and xylene is set by the prevailing gasoline price where they are used as non-oxygenate octane boosters (i.e. alternatives to ethanol or MTBE). Mixed xylene also finds use to produce the important chemical intermediates *para*-xylene (for the production of polyester, PTA), *ortho*-xylene (for the production of phthalates) and ethylbenzene (considered as a xylene and used for the production of styrene[9]). These uses for xylene result in a slight premium over toluene which has no major chemical uses other than for the production of benzene and xylene.

The high concentration of these components makes separation of the BTX an attractive proposition and many naphtha cracking operations now separate BTX rather than produce pyrolysis gasoline.

The other components are C_{5+} olefins and dienes and in particular *cyclo*-pentadiene which easily dimerises to a C_{10} compound (di-*cyclo*-pentadiene). As well as a strong odour, these materials readily polymerise to form gum in the gasoline and the raw pyrolysis gasoline is usually hydro-treated prior to use. In some cases, these C_5 dienes are extracted and used to form low melting resins. One approach to upgrading the pyrolysis gasoline stream is shown in Figure 5.5[10].

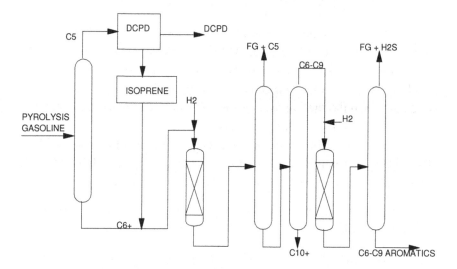

Figure 5.5: Upgrading pyrolysis gasoline

In this approach, pyrolysis gasoline first enters a C_5/C_{6+} splitter which passes the C_5 fraction to a di-*cyclo*-pentadiene unit which dimerises the *cyclo*-pentadiene in the C_5 stream and the dimer is extracted. Excess C_5 is returned to the system via an isoprene extraction unit. The mixture is then hydrogenated and olefins are saturated to paraffins.

A distillation column removes the C_5 and cracked gases to a fuel gas stream. The C_5 stream (now devoid of olefins) can be returned to the cracking furnace. A third column then separates the C_{10+} products (mainly aromatic). A second hydrogenation unit has the duty to remove sulphur and nitrogen from the streams prior to the final distillation to produce an aromatic stream.

A major portion of the world's BTX is made by naphtha reforming. The technology and economics of this route is well reported in petroleum refinery handbooks[11]. Often this route uses extractive distillation to extract aromatics prior to distillation[12]. Reforming operations are often integrated with ethylene cracking operations to maximise benzene production from reformate and pyrolysis gasoline[13].

Fraction Boiling over 190°C; Pyrolysis Fuel Oil

The higher boiling fraction from the cracking of liquid feedstock is generally referred to as pyrolysis fuel oil. Large volumes are produced by cracking gas oil and residual fuel oils. Pyrolysis fuel oil has a greater tendency to form coke than conventional fuel oil and is generally poorer in quality than the fuel oil used as the feedstock. The main use of the fuel oil is as a fuel in the cracking operation.

If the fuel oil uses as the feedstock is low in sulphur, i.e. LSWR, then the pyrolysis fuel oil produced will also be low in sulphur and this makes the product attractive for the production of carbon black.

Storage of LPG

Most liquid feedstock and chemicals such as naphtha or benzene are stored in above ground steel tanks. In order to prevent the formation of explosive mixtures in the tank ullage, some organisations use floating roof tanks, particularly favoured for large tanks in the refining industry, whereas others use inert gas blanket of nitrogen, particularly favoured for small tanks in the chemicals industry.

The main concern is the storage of liquefied petroleum gas (LPG) which in this context can be propane and butane feedstock or product, or ethylene and propylene. The storage of LPG is costly and there are various technologies which depend on the amount of material to be stored.

LPG has been the main source of fuel for some of the petrochemical industries worst fires and explosions. As a consequence the safety measures and regulations concerning LPG storage are extensive and a full discussion of these is beyond the scope of this work.

Storage Tanks

Small volumes of LPG are stored in large cylindrical tanks often referred to as bullets, with typical volumes <100cm.

Figure 5.6 shows a typical above ground spherical storage in which the LPG is stored under pressure. These tanks are common for relatively small volumes of LPG (500 -2,000cm, typically 1,200cm).

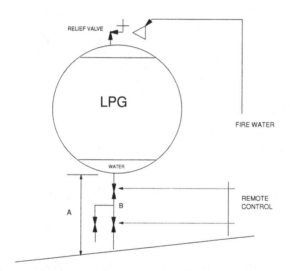

Figure 5.6: Typical layout for a LPG storage tank

Because of a major fire and explosion at Feyzin, France 1966[14] there are extensive regulations concerning the construction and operation of such pressure spheres. The following are some of the general issues to be considered.

The sphere is held at sufficient height (A) to allow easy access to all of the control and operational valves (B), which in general operation are remotely controlled. In normal operation, water contaminating the LPG collects in the base of the vessel which is periodically drawn off. The ground under the sphere slopes so that any leakage of LPG flows away from the area and cannot collect under the sphere. The sphere is equipped with a relief valve, lagging and water dousing in case of fire.

For larger volumes of LPG storage, some organisations use cryogenic storage tanks which are constructed in a similar manner to tanks for the storage of LNG (liquefied natural gas)[15].

LPG Storage in Rock Caverns

For larger volumes of LPG rock caverns can be used. The general layout is shown in Figure 5.7.

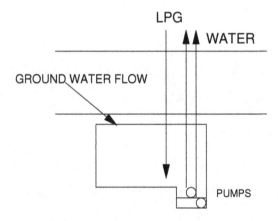

Figure 5.7: Typical layout for a rock cavern

The cavern is constructed by mining out a porous mineral such as limestone, shale or chalk well below the water table. The roof and sides of the cavern are supported by multiple plastic coatings to prevent rock falls. The cavern is sealed hydro-dynamically with the pressure of water entering the cavern balancing the LPG pressure. A well in the lowest part of the cavern contains a submerged pump which pumps excess water to the surface. LPG is removed by pumps higher in the well.

A variation on this is the refrigerated cavern where the ground water around the cavern is frozen and refrigeration plant on the surface has to be provided to support this. It may take 5 years to freeze the ground.

LPG Storage in Salt Caverns

Salt caverns can contain up to 300,000 cm of LPG. Salt caverns were patented in 1916 and have been widely used since the 1950s. They offer absolute tightness and construction from the surface. The general layout is shown in Figure 5.8.

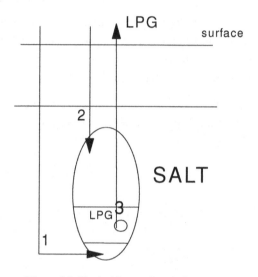

Figure 5.8: Typical layout for a salt cavern

A suitably thick salt layer is located at sufficient depth below the surface. This layer is drilled and water injected and brine extracted which excavates a cavern. Several stages are usually involved. Sometimes this brine is used for caustic-chlorine production.

In operation, brine is injected to the base of the cavern (1) to balance the gas pressure (2). LPG is pumped out of the cavern by the assistance of a submerged pump (3).

Many older established petrochemical complexes are located near such suitable salt fields. This allows the operation to store at relatively modest cost large volumes of ethylene to cover shut-down periods.

Transport

There are three ways to transport chemicals over long distances to a user:

- By land transport in trucks or railcars. This is used if no other method exists for transporting large volumes of feed or product. This is an expensive method and is primarily used as the last link in a supply chain in moving higher valued products to an end user and is not discussed further.
- By ship in a range of carriers of different types and sizes. This is particularly relevant to the international trade in chemicals.
- By pipeline when sufficient volumes of product or feed are to be moved. This is usually used when large numbers of producers and users are in a region (e.g. Western Europe).

This section discusses the economics of chemical transport by ship and pipeline. An overview of the world's shipping fleet for transporting the products of interest is given.

Shipping Fleets

At present, contract shippers conduct most shipping of liquids (chemical and oil derivatives). The merchant fleet is extensive and there is a variety of contracts available for the regular movement of feed and product.

Large amounts of liquid product can be moved in large tankers (>125,000 tonnes). These are generally referred to as dirty cargoes because the product transported is crude oil and residual fuel oil. Generally this fleet is unsuitable to transport chemicals, even if sufficient volumes are available. For the transport of large volumes of chemicals dedicated ships may be required. This may be provided as a contract arrangement.

By contrast the clean cargo fleet (chemicals, naphtha, gasoline) has a wide range of vessel sizes available (10,000 tonnes to over 100,000 tonnes). Transport fuels are typically moved in loads (parcels) of about 80,000 tonnes at a cost of typically $10/tonne (about $0.2/GJ) whereas most chemicals use smaller ships and costs are higher, typically $25-30/tonne, or $1 - 1.5/GJ for chemical methanol.

An important point to note is that contract shipping offers financial advantages over owner operated and dedicated fleets (such as those used to transport LNG), but the contact price is dependent on the vagaries of the shipping market which is both cyclic and seasonal.

For LPG and other liquefied hydrocarbon chemicals such as ethylene, a very large contract merchant fleet is available, although this is dominated by a small number of key players. The available fleet typically moves product at about 30 - 40,000 tonne parcels at a typical cost of $30 - 40/tonne, about $0.6 - 0.8/GJ. However, there are larger ships available (75,000 tonnes).

The cost of contracts is very dependent on business cycles and the season (large LPG demand coinciding with the northern winter). In order to smooth out the costs (from the ship owners perspective) most of the fleet is capable of transporting ammonia and other chemicals as well as LPG cargoes. Thus shipping costs also become influenced by the seasonal nature of ammonia (fertiliser) demand, especially for the US corn market. Table 5.4 gives an overview of the merchant fleet available for the transport of chemicals and fuels.

Table 5.4: Comparison of Transport Fleets for Shipping Chemicals and Fuels

Fleet	LNG	LPG	CHEMICAL	CLEAN FUELS	CRUDE OIL
Products shipped	LNG only	LPG ammonia, chemicals	liquid chemicals	naphtha, gasoline, gas oil	crude oils, fuel oil
Size (tonnes)	90,000	10,000 - 75,000	10,000 - 40,000	60,000 - 120,000	> 120,000
Ship types	Cryogenic	Cryogenic and pressure	Sealed tanks	Sealed tanks	Sealed tanks
Fleet	Dedicated	Contract	Contact	Contact and dedicated	Contract and dedicated
Cost variation	Fixed	Seasonal	Business cycle	Business cycle	Business cycle

Solids Transport

The large scale transport of solids is conducted in large ocean going ships and barges with relatively simple off-loading and on-loading machinery. For example for coal, trans-oceanic transport costs are relatively low. Typical intercontinental costs are $10/t (Australia - Japan) or about $0.33/GJ. For smaller parcels of solids – resins etc. – the world large container fleet can be used.

Estimation of Chemical Shipping Costs

The shipping of chemical methanol is used to illustrate the underlying costs structure of feedstock and chemical shipping. Methanol is liquid under ambient conditions and can be shipped like many other chemicals in closed tanked vessels.

Following the methodology used in this work, two options for methanol transport are considered. The cost basis is based on studies for the transport of methanol as an alternative to gasoline performed by US Department of Energy[16]. The first is for a conventional medium sized tanker of 40,000 t and a very large tanker of 250,000 t. The variation in shipping cost with distance is illustrated in Figure 5.9 and the statistics are given in Table 5.5.

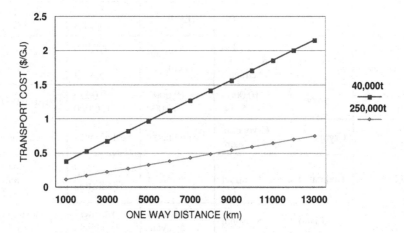

Figure 5.9: Typical shipping costs for medium and large cargoes

Table 5.5: Statistics for Shipping Methanol

Ship Capacity	DWT	40000	250000
	GJ	908000	5675000
Days/year		350	350
One Way Distance	km	5000	5000
Speed	knots	12	12
Sailing time	days	9.37	9.37
Turn around time	h	24	24
One way trips/year		33.74	33.74
Sailing days/year		316.26	316.26
Port calls/year		33.74	33.74
Days in port/year		33.74	33.74
Capital Cost	MM$	37.81	106.56
ROC (15y, 10% DCF)	%	15.19%	15.19%
Annual Capital Costs	MM$/y	5.74	16.81
Labour	MM$/y	2.86	3.76
Fuel	t/day	20	30
Fuel Costs	MM$/y	0.95	1.42
Port Fees/station	$	60000	80000
Port Charges	MM$/y	2.02	2.7
Maintenance	%Capex	4%	4%
	MM$/y	1.51	4.26
Insurances	% Opex	15%	15%
	MM$/y	1.10	1.82
Misc (victualing etc)	%Opex	10%	10%
	MM$/y	0.73	1.21
Total OPEX	MM$/y	9.18	15.18
TOTAL COSTS	MM$/y	14.92	31.36
Quantity Shipped	t/y	674748	4217172
	PJ/y	15.3	95.7
Shipping Cost	$/t	22.11	7.44
	$/GJ	0.97	0.33

LPG Shipping

Many chemicals are liquefied by pressure and transported in tankers similar to the LPG shipping fleet. Transport cost is therefore similar to that of LPG or ammonia, which are transported in either pressurised or refrigerated vessels with costs intermediate between liquids and specialised LNG carriers. LPG shipping costs are seasonal and dependent on the business cycle. Typical costs for the spot carriage cost of LPG cargoes are illustrated in Figure 5.10[17] for three sizes of carrier.

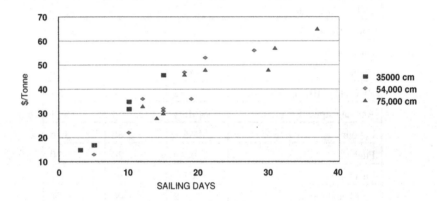

Figure 5.10: LPG shipping costs

This figure shows that as the capacity of the ship rises, then the spot cargo cost falls for a given distance. Interestingly, the cost variation can be accounted for by scaling with an exponent of 0.65; namely:

$$\text{Cost}[1]/\text{Cost}[2] = \{\text{Capacity}[1]/\text{Capacity}[2]\}^{0.65}$$

The result of this normalisation is illustrated in Figure 5.11. The trend-line has the formula:

$$\text{Transport Cost (\$/t)} = 1.64*\text{Sailing Days} + 4.21$$

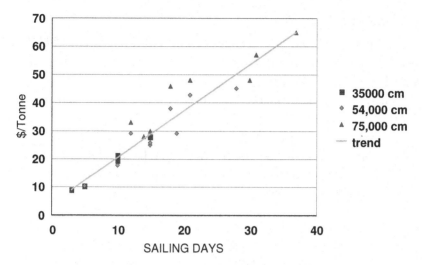

Figure 5.11: LPG shipping cost normalised to 75,000cm vessel

Chemical Pipelines

The movement of ethylene by pipeline to interconnect producers and users is practiced in Europe and the USA. Propylene pipelines are also under consideration.

Despite their simplicity, pipelines are highly capital intensive. Not only is pipe laying costly, the cost is split roughly evenly between materials and labour, but provision of compression stations - necessary for mass transport over long distances - can contribute 40% of finally installed capital. Once established, the operations of a pipeline system can cost 5% of the fixed capital per annum.

The capital cost of a pipeline depends upon such factors as pipe diameter, distance and the amount of compression required. Undersea pipelines cost about double land-based pipelines. Operating costs reflect labour charges and fuel usage in compression if required.

As a rule of thumb, a capital cost of $1 MM/km for land-based pipelines and $2 MM/km for undersea pipelines can be taken as a guide to the capital costs of new pipelines.

LPG transport by pipelines.

Notably, natural gas pipelines can be used to transport LPG. LPG is left in the gas stream in excess of that required by the end users of the gas. The excess is stripped out of the gas further down the line in a straddle-plant. Straddle-plants have been built in Alaska and Western Australia[18].

The use of straddle plants enables the cost of LPG transport to be reduced to the level of that more typical for large scale gas transport which is typically $1/GJ for a pipeline of 1000 km distance. In terms of LPG this is about $50/tonne.

---

[1] Authors analysis of market data reported by *European Chemical News*.

[2] Green oil is a mixture of oligomers and partially hydrogenated oligomers of vinyl acetylene, D. Seddon unpublished results.

[3] Allene is used as an intermediate in the pharmaceuticals industry and *cyclo*-propane is used as an anaesthetic but both are usually directly synthesised for these purposes.

[4] Uhde, "Petrochemical Processes 2005", *Hydrocarbon Processing,* CD ROM, p. 119

[5] Axens, "Petrochemical Processes 2005", *Hydrocarbon Processing,* CD ROM, p. 44.

[6] BASF-AG/ABB Lummus Global Process described in "Petrochemical Processes 2005" *Hydrocarbon Processing* and B. Heida, G. Bohner and K. Kindler, *Hydrocarbon Processing,* Mar. 2002, p. 50B; other processes described *in Petrochemical Processes* include UOP KLP Process which involves selective hydrogenation of acetylenes; "Petrochemical Processes 2005", *Hydrocarbon Processing,* ,CD ROM, p. 38-41.

[7] Anon., *Oil & Gas Journal,* Jan 22, 2001, p. 55

[8] See also CMA study reported in *Oil & Gas Journal,* Mar. 21, 2005, p. 50

[9] Anon., *Oil & Gas Journal,* Feb. 24, 2003, p. 64

[10] Axens technology described in "Petrochemical Processes 2005", *Hydrocarbon Processing*

[11] R. E. Maples, "Petroleum Refinery Process Economics", PennWell, Tulsa, Oklahoma, 1993; J. H. Gary and G. E. Handwerk, "Petroleum Refining – Technology and Economics", 4th Edition, Marcel Dekker, 2001 and "Petrochemical Processes 2005", *Hydrocarbon Processing,* CD ROM, p. 32-37

[12] K. Kolmetz, M. Chuba, R. Desai, J. Gray, A. W. Sloley, *Oil & Gas Journal,* Oct 13, 2003, p. 60; Anon., Uhde Process described in "Refining Processes 2008" *Hydrocarbon Processing,* Sep. 2008, p. 62 and "Petrochemical Processes 2005", *Hydrocarbon Processing,* ,CD ROM, p. 23-27

[13] D. Netzer and O.J. Ghalayini, *Hydrocarbon Processing,* Apr. 2002, p. 71

[14] T. A. Kletz, "What Went Wrong, Case Histories of Process Plant Disasters", 4th Edition, Gulf Publishing, Houston Texas, 1998

[15] LNG storage is discussed in D. Seddon, "Gas Usage and Value," PennWell, Tulsa Oklahoma, 2006

[16] "Assessment of Costs and Benefits of Flexible and Alternative Fuel Use in the U.S. Transportation Sector – Technical Report Three: Methanol Production and Transportation Costs", US Department of Energy DOE/PE-0093

[17] Details from "Waterbourne LPG" 21/06/2001

[18] J. Hawkins, *Oil & Gas Journal*, December 16, 2002, p. 46

CHAPTER 6

CARBON DIOXIDE EMISSIONS

In the petrochemical industry all of the hydrocarbon waste products can be used or recycled in someway, including use as fuel. The principal waste products are water and carbon dioxide. Water is collected, cleaned and recycled for cooling purposes. This leaves carbon dioxide as the major emission in most chemical operations.

At the time of writing, there is increasing concern in many jurisdictions about the emissions of carbon dioxide. This stems from the belief that the accumulated emissions since the advent of the industrial revolution are causing a change in the climate. In order to change the climate it is proposed to limit carbon dioxide emissions. Several methods have been proposed all resulting on a charge to carbon emitting industries. These include:

- Carbon geo-sequestration where all or part of carbon dioxide is captured and stored in deep geological formations.
- A cap and trade mechanism where emitting industries are required to purchase permits for part or all of the carbon emissions.
- A tax on the amount of carbon emissions.

Carbon Geo-Sequestration

Prior to geo-sequestration, the carbon dioxide has to be available in a concentrated form so that it can be compressed and liquefied prior to disposal. From the standpoint of the petrochemical industry, there are two types of emissions which we need to consider. The first type is when carbon dioxide is extracted from a process stream by typically dissolving the carbon dioxide in a solvent. Here the carbon dioxide is available as a concentrated stream when the solvent is regenerated. The second type is associated with fuel burning in the presence of air in furnaces and process heaters. Here the carbon dioxide is emitted diluted in air and excess nitrogen from the furnace or heater stack.

Removal of Carbon Dioxide from Process Streams

Carbon dioxide is produced in petrochemical process streams by reactions with oxygenates (mainly oxygen or water). In steam cracking, hydrocarbons (e.g. methane) and carbon react with steam, forming initially carbon monoxide which is then converted into carbon dioxide by the water-gas-shift reaction:

$$CH_4 + H_2O = CO + H_2O$$

$$C + H_2O = CO + H_2$$

and

$$CO + H_2O = CO_2 + H_2$$

After cooling and removing liquids, the cracked gases are passed to an acid gas plant which has the duty to remove carbon dioxide and any sulphur, which is now present as hydrogen sulphide from the cracked hydrocarbon gases. A typical layout for carbon dioxide removal is shown in Figure 6.1.

Gases containing carbon dioxide enter the bottom of an absorption tower and ascend against the flow of a descending solvent which preferentially absorbs carbon dioxide. The cracked gases, devoid of carbon dioxide, exit the top of the column. The carbon dioxide rich solvent exits the bottom of the tower and is passed to a regenerating column where typically the solvent is boiled to expel the carbon dioxide and regenerate the solvent which is passed back to the absorption tower.

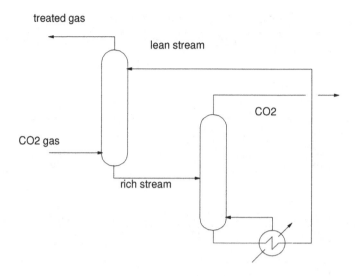

Figure 6.1: Carbon dioxide absorption

Carbon dioxide is also produced as a by-product to oxidation processes. For example, as the by-product to the production of ethylene oxide by oxidation of ethylene with supported silver catalysts:

$$2C_2H_4 + O_2 = 2C_2H_4.O$$

$$C_2H_4 + 4O_2 = 2CO_2 + 2H_2O$$

A similar process can be used to separate the carbon dioxide from unconverted ethylene and product.

Solvent stripping plants of this general type are available in many variants. A common technique for removing carbon dioxide is to wash it with a physical solvent. It is relatively simple but requires a large solvent flow. The solubility of the various components in the solvents are roughly proportional to their boiling point as[1]:

Log [mole fraction of component (1)] α Tb (K)

The boiling points of the non hydrocarbon components of interest are shown in Table 6.1. The normal boiling point for carbon dioxide is

Petrochemical Economics

Table 6.1: Boiling Point of Components of Interest

Gas	Boiling Point (Tb) Kelvin	Comments
H_2	20	
N_2	77	
CO	81	
A	87	
CH_4	112	
$(NO)_2$	122	mp. 112K
CO_2	(175)	Sublimes at 195K, acidic
HCl	188	Acidic
H_2S	213	Acidic
COS	223	
NH_3	240	
SO_2	263	Acidic
HCN	299	Acidic
H_2O	373	mp. 273

extrapolated. If carbon dioxide is the key gas of interest then designing the operation to remove carbon dioxide will result in removal of the higher boiling impurities. This includes hydrocarbons higher than methane which means that the carbon dioxide extraction is usually applied after hydrocarbons of interest have been separated.

Furthermore, the relative solubilities are determined by differences in the boiling points:

$$\text{Log [solubility gas (1)/solubility gas (2)]} = Tb(1) - Tb(2)$$

This can be used to absorb and separate carbon dioxide and hydrogen sulphide. There are several different solvents commonly in use. The properties for carbon dioxide absorption for some solvents of interest are summarized in Table 6.2.

Although the Selexol solvent is an almost ideal solvent for carbon dioxide, there is very little difference between methanol, propylene carbonate and Selexol in terms of mass or solvent volume required. This means in practice there may be little to choose between these alternative solvents.

Table 6.2: Relative Capacity for CO_2 Uptake of Various Solvents

	Water	Methanol	Propylene Carbonate	Propylene glycol dimethyl ether
Technology		Rectisol		Selexol
Molecular Weight	18	32	102	280
Relative CO_2 solubility	2%	12%	35%	100%
Relative CO_2 per gm bar	1	3.6	3.1	3.3
Relative CO_2 capacity per m^3 bar	1	2.8	3.7	3.3

One discriminator is how the solvent handles water. Many solvents, e.g. methanol, are miscible with water. Absorbed water can be removed in a side stream which distils the water from the solvent and hence controls the water content in the main circuit. Some solvents, such as propylene carbonate, decompose when heated with water to 100°C and are therefore unsuitable for treating water wet gases.

Although the common method for degassing solvent is by counter-current flow of gas, concern with capturing carbon dioxide for geo-sequestration purposes may lead to a preference for solvents that can be easily degassed by boiling. This may give a preference to lower boiling solvents such as methanol or N-methylpyrolidone (NMP).

Carbon Dioxide Extraction with Solid Absorbents

Solid absorbents can also be used for removing unwanted materials from gas streams. These can be accomplished by activated charcoal, molecular sieves and increasingly often by membranes[2]. Surprisingly, the relative selectivity of these absorbents is again proportional to the relative boiling points. To a large extent these materials condense liquids by reducing their vapour pressure as a result of very high negative hydraulic pressures exerted by surface tension of the liquid in very fine capillaries. Again, focussing on carbon dioxide removal generally removes higher boiling materials. Activated charcoal has now largely been replaced by molecular sieves (e.g. Polybed PSA Process).

Absorbent processes generally work with a high pressure (20–50 bar), with desorbing and bed regeneration at low pressure (1 bar), hence adiabatic operation. The process can be used to absorb carbon dioxide, carbon monoxide and methane from hydrogen streams.

The solubility of gases in membranes is again proportional to the gas boiling point. After dissolution, the absorbed gas diffuses through the membrane to the lower pressure, lean gas side at rates which are inversely proportional to the square root of the gas molecular weight. Multiplying relative solubility by the relative diffusion constant gives relative permeability. Carbon dioxide is 10 times more permeable than hydrogen and water is 1,000 times that of hydrogen. However, there is a problem with the ability of the gases to "wet" the surfaces and this can greatly reduce the relative permeability from the theoretical values. The consequence of this is that membranes tend to be used for hydrogen recovery processes rather than for carbon dioxide extraction *per say*.

Carbon Dioxide Absorption with Chemical Absorbents

Carbon dioxide and some other gases which require removal from process streams are acidic. In order to improve the absorptive capacity of physical solvents, basic (alkaline) chemicals are added. The choice of chemical additive is determined by the ability to pick up the component of interest and to be able to release it again in the de-sorber. The absorption must be reversible and preferably by dropping the pressure.

For carbon dioxide absorption, the heat of absorption could provide the heat required to desorb the carbon dioxide. The system would be adiabatic. However, the only effective solvent is the alkaline carbonate to bicarbonate reaction – The Benfield Process[3].

$$CO_2 + CO_2^= \ = \ 2HCO_3^-$$

And for hydrogen sulphide:

$$H_2S + CO_2^= \ = \ HCO_3^- + HS^-$$

The problem is that effective operation is at about 100°C and the only possible stripping gas which would not abstract a great deal of heat from the desorber unit is steam, and hence this process has a very high steam demand. The high steam demand lowers the attraction of this process relative to the newer processes using amines.

The main reagents used for carbon dioxide and hydrogen sulphide removal are based on alkanolamines. These form amine carbamates:

$$CO_2 + 2R_1R_2NH = R_1R_2NCOO^- + R_1R_2NH_2^+$$

For MEA and DEA the equilibrium lies to the right so that regeneration (like Benfield) is conducted at 100°C. Unfortunately, amines which would operate at lower temperatures have lower kinetic factors for carbon dioxide uptake.

The steam requirement in the stripping section of these chemical absorbent processes is the main disadvantage of these systems relative to the physical sorption routes. However, a major advantage is that losses of non acidic gases – hydrocarbons, hydrogen, carbon monoxide etc. – are much lower because the absorption of these components is not influenced by the chemicals used.

Because chemical absorbents can be used at low pressure they are considered the optimum method for extracting the carbon dioxide from the dilute gases.

Removal of Carbon Dioxide from Flue Gas

Most of the carbon dioxide in the petrochemical industry is emitted in flue gases as a result of burning fuel oil and fuel gas. The basic problem can be illustrated by considering the combustion of a fuel gas considered as methane.

If methane is burned in air the stoichiometric equation is:

$$CH_4 + 2O_2 + 8N_2 = CO_2 + 2H_2O + 8N_2$$

On a dry basis, the concentration of carbon dioxide nitrogen in the flue gases will be 11.1% (1/9). However, in practice excess air is used to avoid incomplete combustion and the emission of soot from the flue stack. Excess air serves to reduce the carbon dioxide concentration.

However, the main problem with excess air is that it introduces oxygen into the flue gas. Typical flue gas concentrations for natural gas and coal combustion in power production are given in Table 6.3.

Table 6.3: Typical Flue Gas Compositions

	NATURAL GAS	COAL
CO_2	2 – 7.5%	9 – 15%
H_2O	9 -10%	6 – 16%
N_2	72 – 73%	70%
O_2	4.5 – 18%	3 – 21.5%

The excess air's introduction of oxygen degrades the principal chemical absorbents (alkanolamines) and increases solvent consumption. This degradation is also exacerbated by the presence of sulphur in the flue gas, e.g. from fuel oil. This is illustrated in Figure 6.2 and Table 6.4 which illustrates process flows and utilities consumption for the reduction of carbon dioxide in a flue gas to below 2% using a standard sorbent (MEA) and a solvent under development (SH amine)[4].

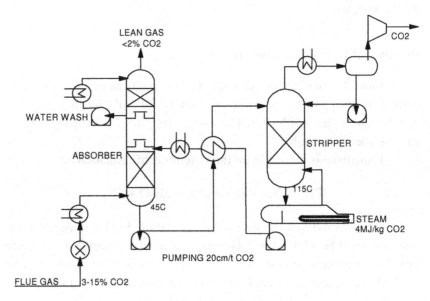

Figure 6.2: Flue gas stripping unit

Table 6.4: Typical Utility and Solvent Consumption in Flue Gas Scrubbing

	MEA	SH amine
Steam for solvent t/t CO2	1.95 to 3.0	1.2
Regeneration GJ/t CO2	4.2 TO 6.5	2.6
Solvent flow cm/t CO2	17 to 25	11
Power (pumps) kWh/t CO2	150 to 300	19.8
Cooling water, cm/t CO2	75 to 165	150
Solvent consumption kg/t CO2	0.45 to 2.0	0.35
SO2 tolerance, ppm	10 to 100	<10

As well as the utility costs shown in Table 6.4, there is the issue of compressing the flue gas to sufficient extent to be able to flow upward through the absorber against the falling solvent. This compression cost is exacerbated by the high levels of nitrogen present.

One of the main concerns with emission reduction is the state of large central base load power facilities using fossil fuels, in particular coal. These emit large volumes of carbon dioxide through flue gases. As a consequence there is extensive research and many demonstration projects aimed at extracting carbon dioxide from flue gases. As well as research on better sorbents, other methods are being proposed: this includes the use of oxygen in place of air in the combustion process.

Strategies for Reducing Carbon Emissions

What is obvious from the data provided above is that the flue gases can still contain 2% of carbon dioxide. Not only will they attract emissions charges but they will compromise the goal of zero emissions. One way of improving the capture of carbon dioxide from the flue gas would be to use oxygen as opposed to air for the fuel combustion process. This will eliminate the very large excess of nitrogen in the flue gas. However, introduction of oxygen would dangerously increase the furnace combustion temperature. This is proposed to be solved by recycling a large volume of the flue gas (which would comprise carbon dioxide and water vapour) to the furnace. The system still leaves unresolved oxygen and sulphur contamination of the flue gas which will lead to degradation of the current class of absorbents.

One way to reduce carbon emissions is to use fuels of lower carbon intensity. The carbon intensity of some fuels of interest is shown in Table 6.5.

Table 6.5: Carbon Intensity of Some Fuels of Interest

FUEL	% CARBON	Tonne Carbon/TJ	Tonne CO2/TJ
Natural Gas	76	14	51.3
LPG	81	16.4	59.4
Naphtha	87	18.2	66
Fuel Oil	89	19.2	69.7
Brown Coal	25	26.2	95
Black Coal	67	24.8	90
Wood	42	25.9	94
Bagasse	26	26.7	96.8

The first group of fuels, natural gas, LPG, naphtha and fuel oil, are those which are typically used in furnace operations in the petrochemical industry. This illustrates that moving from fuel oil to natural gas can achieve significant reductions in the carbon emission of a site. However, it must be remembered that on a global (cradle to grave) basis this may overestimate the benefit as these figures ignore the carbon emission in production of the fuel. This can be quite substantial for natural gas when the raw gas in the field is contaminated with carbon dioxide; many fields contain 30% (mass basis) or more carbon dioxide which is stripped from the raw gas in gas plant operations in order to produce gas of a quality that can be piped (typically <2% vol.) carbon dioxide.

A second group is coal which is seen to have much higher carbon intensity than the liquid or gaseous fuels. Coal is often used by power generation operations associated with petrochemical operations. This power is often purchased from a third party and on a global basis should be counted if power is imported. However, at the present time this type of carbon counting is not demanded by many jurisdictions leaving the option for a petrochemical operation to import power from coal facilities without incurring the carbon cost. This raises the possibility of a petrochemical operation converting to oxygen fired furnaces and using imported power thereby "reducing" the carbon emissions of the site.

The third group of fuels in Table 6.5, wood and bagasse[5], are representative of renewable fuels. Although these fuels have a high carbon emission intensity they are considered benign to the climate change argument and do not count to emissions. One problem with these fuels for the substitution of black coal is that they have a high water content and degrade the thermal efficiency of the power generation process, i.e. less power is produced.

To this list should be added renewable ethanol (i.e. from crops) and biodiesel which are more easily substituted for liquid fuels. These can be used to displace fossil fuels used in the furnace and hence lower the carbon emission. Unfortunately, in the opinion of the author, these fuels often involve major ecological impacts elsewhere such as displacement of food producing agricultural land (ethanol) or destruction of native forests (biodiesel) for the production of a favoured feedstock – palm oil.

One way to improve carbon emissions and overall efficiency is to ensure that all furnace operations employ efficient heat recovery from the flue gas. Ideally the flue gas should be cooled in order to recover the heat of condensation of the water produced in the combustion process.

Geo-Sequestration

For geo-sequestration of carbon dioxide in flue gas it may be necessary to first remove sulphur so as to protect the carbon dioxide solvent. The flow from the flue system has to be modified to include an additional plant prior to compression, carbon dioxide and geo-sequestration as illustrated by Figure 6.3.

The first operation is to remove residual heat from the flue gas and potentially generate steam which can be used in the carbon dioxide scrubber (MEA). Next, the flue gas is treated with limestone to reduce the sulphur content of the gas stream. The stream is then compressed to a level sufficient for the scrubbing operation. The carbon dioxide depleted flue gas is passed to the stack and the carbon dioxide extracted is compressed prior to injection in the well.

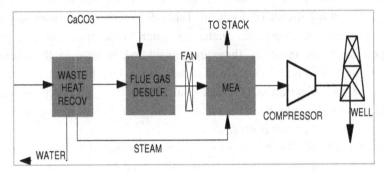

Figure 6.3: Addition process plant required for flue gas CO_2 scrubbing

The level of compression of carbon dioxide required is dependent on the disposal option but can generally be said to be in the range of 150–180 bar for disposal in saline aquifers and depleted oil reservoirs. Disposal in coal measures may require less compression (80–100 bar) and deep sea trenches more (250–300 bar). High capacity carbon dioxide injection plants are complex and require multi-stage compression steps[6]. This amount of compression requires significant levels of power, this has been estimated by Saxena and Flintoff[7] and summarised in Table 6.6 for 500,000 t/y carbon dioxide.

Taking into account that a large naphtha cracking operation producing 1 million tonnes of ethylene will emit about 3 million tonnes of carbon dioxide, approximately 40MW of compression power will be required.

Table 6.6: Estimates of Power Required for Carbon Dioxide Compression

t/y CO2	500,000
cm/h CO2	39,411
Suction Pressure (kPa)	115
Discharge Pressure (kPa)	18,000
Power Required kW	6,610

At the time of writing, the major unresolved issue is the choice of a site that can store the quantum of carbon dioxide and guarantee its safe disposal for many hundreds of years. It is not clear if depleted oil and gas reservoirs can do this and it may require major exploration for suitable sites. The problem is further exacerbated by the remoteness of suitable sites from major producing facilities, often several hundred kilometres[8]. This will incur a further significant pipeline and transport cost.

The Cost of Carbon Geo-sequestration

Some guidance to the cost of carbon geo-sequestration is obtained by consideration of the costs of using carbon dioxide in enhanced oil recovery schemes[9]. This is in the region of $20-25/tonne of carbon dioxide after purchase of the gas for this duty.

Taking this as a cost for the disposal part, we can guess that extraction of the carbon dioxide (as illustrated in Figure 6.3) and compression would more than double this amount. This gives a ball-park estimate of $50/tonne at the least. This cost will be further significantly increased if the disposal site is remote from the facility. At the time of writing it is worth noting that emission credits are a fraction of this cost.

In order to give an informative account of the likely cost of carbon abatement of the various petrochemical operations, the carbon emissions have been estimated and a fixed variable relationship developed with carbon dioxide disposal cost as the variable. This cost can be either the cost of geo-sequestration facilities, the cost of purchasing emissions certificates or carbon tax.

A final note is that some facilities inject both hydrogen sulphide and carbon dioxide from acid gas plants[10] making it feasible to dispose of both gases simultaneously.

---

[1] This analysis follows work of S. P. S. Andrews at ICI Billingham (private communication)

[2] G. Blizzard, D. Paro and K. Hornback, *Oil & Gas Journal*, Apr. 11, 2005, p. 48; A. Callison, G. Davidson, *ibid.*, May 28, 2007, p. 56; J. Marquez, M. Brantana, *ibid.*, Jul. 24, 2006

[3] "The Benfield Process for Acid Gas Removal", R. K. Bartoo in "Acid and Sour Gas Treating Processes", S. A. Newman (ed.), Gulf Publishing, Houston, Texas,1985

[4] Saxena and Flintoff, Hydrocarbon Processing, December 2006, p. 57

[5] Sugar cane waste

[6] S. Ariyapadi, J. Strickland, J. Rios, *Oil & Gas Journal*, Sep. 4, 2006, p. 74

[7] M. N. Saxena, W. Flintoff, *Hydrocarbon Processing*, Dec. 2006, p. 57

[8] G. Moritis, *Oil & Gas Journal*, Mar. 3, 3003, p. 39

[9] M. K.Dubois, A. P. Byrnes, R. E. Pancake, P. G. Willhite, L. G. Schoeling, *Oil & Gas Journal*, June 5, 2000, p. 37; G. Moritis, *ibid.*, May 14, 2001, p. 68; Anon., *ibid.*, May 17, 2004, p. 48

[10] S. G. Jones, D. R. Rosa, J.E. Johnson, *Oil & Gas Journal*, Mar. 1, 2004, p. 54; *idem.*, Mar. 8, 2004, p. 45

ECONOMIC ANALYSIS

CHAPTER 7

ETHANE CRACKING

This chapter describes the basic process of ethane steam cracking operations to produce ethylene and the integration with downstream operations. The approach to economic analysis for various types of ethane cracking operations is described and the economic analysis for ethane cracking in a standardised approach is developed. The production of olefins from other feed stocks and the economics of production are developed in later chapters[1].

Ethane cracking is conducted across the world. The scales of operation range from the smallest, less than 50,000 t/y, when small amounts of ethylene is required, for example for a stand-alone styrene plant, to the largest ethylene production operations of over 1 million tonnes of ethylene. The block flow layout for a small stand-alone ethane cracking operation is illustrated in Figure 7.1.

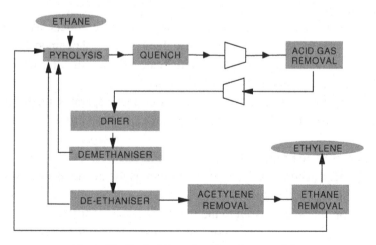

Figure 7.1: Typical flow-sheet for ethane cracking

125

Ethane enters the pyrolysis section, which comprises a series of cracking furnaces. The ethane is heated as quickly as possible to the cracking temperature and maintained at this temperature for the minimum residence time. In order to lower the hydrocarbon partial pressure and mitigate coke forming in the pyrolysis tubes, steam is added to the ethane prior to entering the pyrolysis section (not shown).

Immediately after cracking, the cracked gases are reduced in temperature as quickly as possible to stop the cracking processes and prevent the cracked gases forming coke. This quenching is often referred to as the transfer-line-heat-exchange (TLE). Excess steam is condensed and the water recycled (not shown). Heat from the TLE is recovered as process steam.

Cracking is best conducted at low pressures, whilst product cleaning and separation is best conducted at higher pressures. The cracked gases are now compressed and passed to an absorber tower where acid gases (carbon dioxide, sulphur compounds) are extracted from the cracked gas stream. The absorbent is regenerated (usually by heating) and the acid gases are passed to other downstream units for disposal as may be necessary.

The gases are then compressed further and passed through a drier to the separation train. The cracked gases now contain only hydrocarbons and hydrogen.

Ethane cracking produces a range of by-products as well as ethylene. However, relative to other feed stocks, the amount of by-products is small and in many small-scale ethane cracking operations these are used as fuel in the pyrolysis furnace.

The first separation unit is a tower that separates methane and hydrogen from the C_{2+} gases (de-methaniser). These are used as fuel in smaller cracking operations, but can be further separated in the larger scale crackers to produce a fuel gas and hydrogen.

The second separation tower, known as the de-ethaniser, separates the C_{3+} fraction from the C_2 cracked gases. In smaller operations the heavier products are passed to the pyrolysis furnace for fuel.

The C_2 cracked gases contain a small amount of acetylene, which is usually removed in an acetylene hydrogenation unit. To accomplish this, a small portion of hydrogen is added (extracted from the de-methaniser off-gas) and the gas mixture passed over a palladium catalyst which selectively removes the acetylene by hydrogenation:

$$C_2H_2 + H_2 = C_2H_4$$

And

$$C_2H_2 + 2H_2 = C_2H_6$$

Acetylene hydrogenation is widely practiced and efficient. However, a "green-oil" which comprises vinyl acetylene oligomers is also produced and in some instances can foul the unit. In large cracking operations, the acetylene may be recovered by absorption processes based on copper salts, which selectively absorb the acetylene.

The gases are now passed to a splitter column which separates the ethane and ethylene. Ethane cracking has a relatively low pass conversion and there are relatively large amounts of ethane present in the ethylene stream. After separation, the ethane is recycled to the feed section where it is cracked to extinction. The ethylene is passed to downstream units for production of other chemicals and resins.

In larger-scale ethane cracking operations, or those integrated into large chemical complexes, the useful by-products can be separated and used. In this instance the pyrolysis furnace is fired by fuel oil. Note that different process licensors have differing approaches to the layout of the unit operations[2]. A typical situation is illustrated in Figure 7.2.

In this configuration, the hydrogen and the methane from the de-methanizer column are split into their component streams. The hydrogen is for use in various downstream processes and the methane is used as a fuel-gas stream. Bottoms from the de-ethaniser are further split into C_3 and C_{4+} stream. The C_3 is treated similarly to the C_2 to produce polymer grade propylene. After removing the C_4 fraction, which is passed to downstream separation units, the heavy components form pyrolysis-gasoline. The latter may be further separated to produce benzene, toluene and xylene.

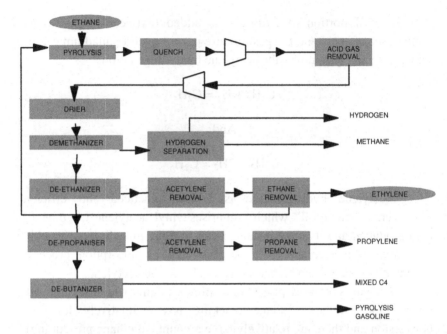

Figure 7.2: Typical flow-sheet for large scale ethane cracking

Feed Requirements and Thermal Efficiency

Table 7.1 presents typical data from a large ethane cracking operation designed to produce 500,000 tonnes per year of ethylene with a flow-sheet similar to that given in Figure 7.2.

In order to produce 500 kt/y of ethylene some 651 kt/y of ethane is required. However, to allow for process upsets and general operational issues, a feed operating allowance of 5.5% is typically added. The pyrolysis requires 330 kt/y fuel oil for the furnace fuel. Again, an operating allowance, this time of 3.5%, is made.

The second column gives the corresponding energy of the feed stock and fuel streams in PJ/y. The total feed and fuel required is 1.027 million tonnes (Mt/y) corresponding to 50.2 PJ/y (HHV basis).

The by-products in producing 500 kt/y ethylene by ethane cracking are 17,000 t/y each of propylene and mixed C_4 stream

Table 7.1: Ethane Cracking in Integrated Operations

	kt/y	PJ/y
INPUTS		
Feedstock	651	33.787
Operating Feed (5.5%)	36	1.801
Fuel	330	14.157
Operating Fuel (3%)	10	0.425
TOTAL	1027	50.170
OUTPUTS		
Ethylene	500	25.150
Propylene	17	0.831
BTD/C4 olefins	17	0.818
Gasoline	11	0.510
Hydrogen	66	9.359
Methane	40	2.220
Propane	0	0.000
Butane	0	0.000
Fuel Oil	0	0.000
TOTAL	651	38.888
THERMAL EFFICIENCY (%)		77.513

comprising butadiene and olefins (BTD/C4), 66,000 t/y hydrogen and 11,000 t/y of pyrolysis gasoline. Moreover, 40,000 t/y methane (fuel gas) is also produced. The total of ethylene and by-products is 651,000 t/y corresponding to the required feed without an operating allowance.

The total energy of the products is 38.9 PJ/y. This gives a total operating efficiency for the plant of 77.5%.

Inspection of the energy available in the by-products indicates that there is ample energy available for operating the pyrolysis furnaces without the need to use any fuel oil (as the flow-sheet in Figure 7.1). This operational method degrades high value propylene and gasoline streams to fuel oil value. This may be justified for smaller operations, but for larger scale operations it is then more economically attractive to extract and sell the propylene and pyrolysis gasoline.

Table 7.2 presents the data for a plant which extracts propylene and pyrolysis gasoline, but recycles the rest of the products. Of the by-products the mixed C_4 stream is recycled to the feed-side of the cracker furnaces, with the hydrogen and methane recycled to the fuel-side. The same quantum of operating allowances for feed and fuel are included in the statistics.

The result of this by-product recycling is to reduce the feed demand and almost eliminate the requirement for fuel oil. Operating allowances are maintained.

Table 7.2: Ethane Cracking with Some By-product Recycle

	kt/y	PJ/y
INPUTS		
Feedstock	635	32.969
Operating Feed (5.5%)	36	1.801
Fuel	60	2.578
Operating Fuel (3%)	10	0.425
TOTAL	741	37.773
OUTPUTS		
Ethylene	500	25.150
Propylene	17	0.831
BTD/C4 olefins	0	0.000
Gasoline	11	0.510
Hydrogen	0	0.000
Methane	0	0.000
Propane	0	0.000
Butane	0	0.000
Fuel Oil	0	0.000
TOTAL	528	26.492

Ethylene Production Costs From Ethane

Base Case Analysis

The procedure adopted is to establish a base case, which is representative of the average operation of interest and then to address the sensitivity of the base case against the key economic variables. The base case is developed around the production of 500,000 t/y ethylene using ethane at a cost of $7.19/GJ ($373.3/t). This ethane price is discussed in an earlier chapter and corresponds to a natural gas price of $6.37/GJ (average US 2007 price) into a suitably large-scale gas plant.

The capital cost of a 500,000 t/y ethane cracker is $718 million (2007). The non-feedstock operating costs are taken as 10% of the capital per annum or $71.8 million per year (MM$/y). If the plant is built in three years on the basis of a 20 year life with a DCF rate of 10%, annual return on capital, as detailed in the Appendix, is 14.4% or $102.8MM/y.

The production economics can be estimated as a function of ethane price using the assumptions:

- Propylene by-product is valued at the 95% value as ethylene ($/t basis).
- The prevailing price of oil (WTI) is set at $70/bbl. This sets the price for fuel oil and pyrolysis gasoline. The latter is priced at a $1/bbl discount to refinery gasoline to allow for its poorer quality relative to regular traded gasoline.
- The other by-products are taken as the 2007 year average or for mixed C4 stream equivalent to that of naphtha.

The two systems described by Tables 7.1 and 7.2 are evaluated. For ease of discussion, the flow-sheet described by Table 7.1 where all possible cracked products are sold at prevailing market prices is referred to as OPEN. The case where some of the product is recycled to feed or fuel (Table 7.2) is referred to as CLOSED.

Setting the ethane price to $7.19/GJ (which corresponds to a gas plant price with gas available at $6.37/GJ) gives the ethylene production cost of $726/tonne for the OPEN system and $869/tonne for the CLOSED system. The cash flows are detailed in Table 7.3.

Table 7.3: Ethane Cracking Cash Flows (MM$/y)

	kt/y	OPEN	CLOSED
OPEX (10% CAPEX)		71.88	71.88
RETURN ON WORKING CAP(10%)		7.04	5.56
RECOVERY (10%DCF, 20y, FACTOR 0.143)		102.79	102.79
INPUTS			
Feedstock		243.00	237.12
Operating Feed (5.5%)		13.36	13.36
Fuel		118.04	21.50
Operating Fuel (3%)		3.54	3.54
TOTAL		377.94	275.52
OUTPUTS			
Ethylene	500		
Propylene	17		
BTD/C4 olefins		10.17	
Gasoline		7.11	7.11
Hydrogen		149.74	
Methane		17.76	
Propane		0.00	
Butane		0.00	
Fuel Oil		0.00	
TOTAL		184.78	7.11
ANNUAL COSTS		559.64	455.74
BYPRODUCT CREDITS		184.78	7.11
UNIT ETHYLENE PRODUCTION COST ($/t)		726.27	869.19

Comparison of the data in the columns of the OPEN and CLOSED cases clearly demonstrates that degrading cracked products to feed or fuel value results in a dramatic increase in the unit production cost.

The break-up of the production costs for the two systems are illustrated in Figures 7.3 and 7.4 for the OPEN and CLOSED systems respectively.

The largest cost item of the open system is the feedstock cost, followed closely by the capital and fuel costs at nearly 20% each. Non feed and fuel operating costs are about 14% of the total unit production costs.

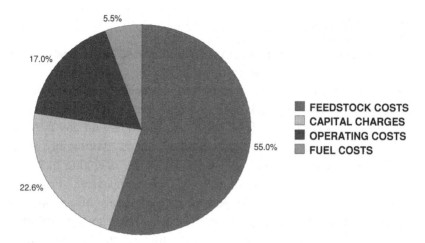

Figure 7.3: Breakdown of ethylene production costs -ethane feed - OPEN system

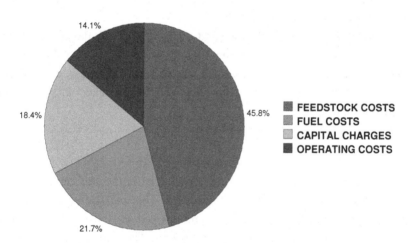

Figure 7.4: Breakdown of ethylene production costs -ethane feed - CLOSED system

For the CLOSED system, after feedstock costs, capital charges are the largest item followed closely by non-feed operating costs. Fuel cost is very much reduced as the unsold by-products are used to displace the fuel import into the plant. Fuel cost is only about 6% of the total. Non feedstock operating costs and fuel costs are about 22% of the total unit costs.

Operating Margin

In some cases, analysis of production economics is done from the standpoint of an operating margin that expresses the difference between the realised product costs and the cost of feedstock. This method takes no account of the non feedstock operating costs (labour, maintenance, etc.) and the cost of capital, both of which are dependent on any particular facility. The method thus expresses the general profitability of the operations in a given situation. Analysis of some operating margins is published by various consultancy service companies[3]. Proprietary furnace models also simulate performance in terms of an operating margin which allows operators to optimise production costs in real-time[4].

Table 7.4 illustrates a typical operating margin calculation. The unit operating margin calculated for the OPEN and CLOSED systems when the value of the feedstock and products are as shown in the table are $749/t and $606/t respectively.

Inspection of the Table and comparison of the OPEN and CLOSED cases demonstrates the loss in profitability as valuable by-products are downgraded to feedstock or fuel value. This is particularly evident for hydrogen, whose by-product credit makes a significant contribution to profit.

Sensitivity to Crude Oil Price

The primary impact of rising oil price is on the cost of fuel oil for cracking and on the value of by-products. This particularly affects the OPEN system where by-products are on-sold or transferred to other downstream operations at world parity prices. The sensitivity of ethylene production cost to oil price in the range $30 to $230/bbl is illustrated in Figure 7.5, which shows the cost of increasing oil price on the OPEN system. At $70/bbl the base production cost is about $726/t and falls to about $666/t with oil at $50/bbl and rises to $964/t with oil at $150/bbl.

For the CLOSED system sensitivity curve has a flat or slightly rising profile because the by-products are used to displace fuel oil and as the price of oil rises the value of the former rises faster than the latter.

Table 7.4: Margin Calculations using Ethane Feedstock

		OPEN		CLOSED	
	$/t	kt/y	MM$/y	kt/y	MM$/y
FEEDSTOCK & FUEL PURCHASES					
Feedstock	373.27	651	243.00	635	237.12
Operating Feed (5.0%)	373.27	36	13.36	36	13.36
Fuel	357.68	330	118.04	60	21.50
Operating Fuel (3%)	357.68	10	3.54	10	3.54
TOTAL		1027	377.94	738	275.52
OUTPUTS					
Ethylene	1278.00	500	639.00	500	639.00
Propylene	1214.00	17	20.64	17	20.64
BTD/C4 olefins	597.95	17	10.17	0	0.00
Gasoline	646.37	11	7.11	11	7.11
Hydrogen	2268.80	66	149.74	0	0.00
Methane	444.00	40	17.76	0	0.00
Propane	561.90	0	0.00	0	0.00
Butane	533.42	0	0.00	0	0.00
Fuel Oil	357.68	0	0.00	0	0.00
TOTAL		651	844.41	528	666.75
MARGIN		MM$	387.56		352.95
		$/t	749.63		606.95
		c/lb	34.00		27.53

The production cost remains at about $870/t over the range of $35 to $100/bbl oil.

Sensitivity to Ethane Price

For ethane feedstock, of most interest is the sensitivity of the production cost to the price of ethane. In many jurisdictions, the ethane price is related to the price of gas. In turn this is related in many parts of

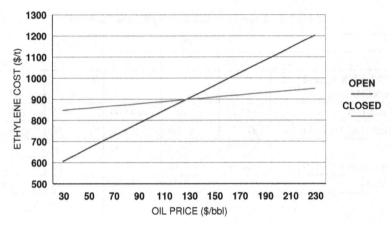

Figure 7.5: Ethylene from ethane – sensitivity to oil price

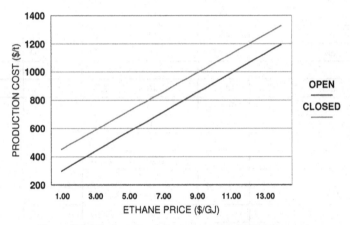

Figure 7.6: Ethylene from ethane – sensitivity to ethane price

the world to the prevailing price of oil. However, in many other parts of the world the price of gas is disconnected from the price of oil and hence ethane is priced on the cost of its extraction. Because the ethane price is not universally linked to oil price, a range of ethylene production costs exists for ethane cracking operations across the world[5]. The sensitivity of the base case to the price of ethane is illustrated in Figure 7.6. In this figure the sensitivity is plotted against the price of ethane in energy units as $/GJ (1GJ is approximately 0.95MMBTU).

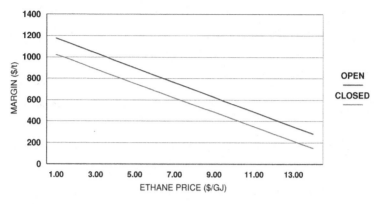

Figure 7.7: Ethylene from ethane – sensitivity of margin to ethane price

The figure illustrates the rise in production cost with the rise of feedstock price for the base case. For the OPEN system (all by-products sold), doubling the price of ethane from the base value (about $7/GJ) results in a production cost rise from about $710/t to about $1200/t. However, more critically, lowering the price of ethane to about $2/GJ results in ethylene production costs below $400/t. For the CLOSED system (some by-products recycled), the production costs line runs parallel to the OPEN system line, but at about $150/t higher. For low price ethane ($2/GJ); the production cost of ethylene is about $500/t. The influence of changing the ethane price on the operating margin is shown in Figure 7.7. This illustrates that as the ethane price rises from $2.50/GJ, roughly corresponding to well head gas of $2/GJ into a gas plant, to $7.50/GJ, roughly corresponding to the opportunity value of leaving the gas in pipeline gas in developed economies at high oil price, the operating margin halves.

Sensitivity to Scale of Production

The sensitivity of ethane production cost to the scale of operation at two representative ethane prices for the OPEN and CLOSED systems are shown in Figure 7.8. The graphs span the range of production capacities from the largest to the smallest ethane cracking operations. The graphs have been derived assuming capital cost vary according to a

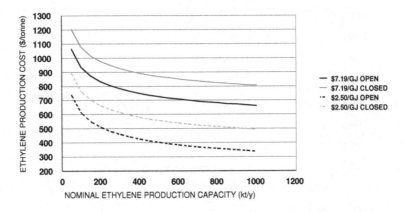

Figure 7.8: Ethylene production cost – impact of scale of operation

power function with an exponent factor of 0.7. Both the OPEN and CLOSED systems are evaluated at a high and a low ethane price ($7.19/GJ and $2.50/GJ).

Relative to the base case with a production capacity 500,000 t/y, production costs fall by $100/t in both the OPEN and CLOSED systems with capacity increases to 1.2 million tonnes per year. Conversely, smaller scale operations see increases in production costs of about $350/t for small scale (50,000 t/y) output.

Small scale operations are widely used to produce small amounts of ethylene for a specific purpose (e.g. styrene). This graph illustrates that high ethane prices are a significant threat to these operations because the cost of ethylene transport from a larger operation (typically $100/t for ship based transport) is lower than the rise in production cost due to the loss of economy of scale.

Comparing the smaller scale operations with access to low ethane prices shows that these are competitive with much larger operations paying high ethane prices.

Production of Polymer

Many cracker operations are integrated into the downstream production of polymers and resins. For ethane cracking this usually means integration into the production of various polyethylene grades.

Integration with polypropylene manufacture can occur in those cases when production is very large, or there is some cracking of heavier material or else additional propylene can be sourced from a local refinery or a propylene specific production operation.

The reason for this integration is that extra product value can be captured with modest additional capital expenditure. This case is especially found in smaller stand-alone cracking operations (CLOSED case) where there is little opportunity to dispose of the minor by-products

Table 7.5: Polymer Production Cost and Operating Margin

	$/tonne	kt/y	MM$/y	MM$/y
CAPEX			$960.11	
OPEX (10% CAPEX)			96.01	
RETURN ON WORKING CAP (10%)			$7.38	
RECOVERY (10%DCF, 20y, FACTOR 0.143)			137.3	
FEEDSTOCK & FUEL PURCHASES				
Feedstock	373.27	635	$237.12	$237.12
Operating Feed (5.5%)	373.27	33	$13.36	$13.36
Fuel	357.68	60	$21.50	$21.50
Operating Fuel (3%)	357.68	10	$3.54	$3.54
TOTAL		738	$275.52	$275.52
OUTPUTS				
Polyethylene	1702.4	500		851.20
Polypropylene	1590.4	17		27.04
Py-Gasoline	646.37	11	7.11	7.11
TOTAL		528	7.11	885.35
ANNUAL COSTS			516.20	378.91
BYPRODUCT CREDITS			7.11	
POLYMER PRODUCTION COST ($/t)			986.84	
MARGIN				506.44
			$/t	978.58
			c/lb	44.43

other than as cracker fuel. The statistics for estimating the production cost and the operating margin are shown in Table 7.5.

For the sake of simplicity if we assume a 100% conversion of olefin into polymer then the production cost of polymer is $986/tonne compared to the production of ethylene in the closed system of $869/t (Table 7.3). With polymer prices at the $1700/tonne level (2007 basis), the operating margin is $978/tonne, somewhat higher than can be found for ethylene production (Table 7.4) and for a more easily handled and sold product.

The impact of ethane price on the operating margin is illustrated in Figure 7.9 and illustrates the erosion of the margin as the price of feedstock rises from the low to high values. At the higher end of the range of ethane prices, the operating margin is reduced to a point that is close to the typical month to month variation in the price of polymer.

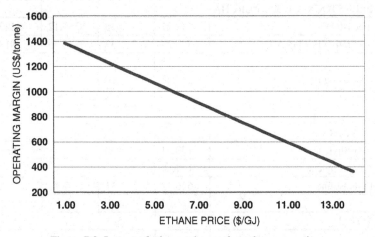

Figure 7.9: Impact of ethane price on the polymer margin

The graph clearly demonstrates the competitive advantage of operations with ethane tied to a low cost gas price (ethane prices <$4/GJ) compared to those operations with ethane linked to the price of oil.

Carbon Emissions from Ethane Cracking

If we assume that all utility steam is raised from heat exchange of hot furnace gases and that excess steam is used to generate all of the power required by the cracking plant, then carbon emissions come from two major sources:

- The fuel used to heat the furnace and drive the cracking reactions and
- Operational losses which result in the flaring of feed stock or product.

The estimated emissions from these two sources in ethane cracking are set out in Table 7.6:

Table 7.6: Estimated Carbon Emissions from Ethane Cracking

	OPEN			CLOSED		
Fuel Used	Mt/y	PJ/y	MtCO2/y	Mt/y	PJ/y	MtCO2/y
Methane					2.22	0.114
LPG						
Fuel Oil	0.34	14.582	1.016	0.07	3.003	0.209
Total	0.34	14.582	1.016	0.07	3.003	0.209
Emissions/t ($C_{2=} + C_{3=}$)			1.966			0.405
Flaring						
Ethane	0.036	1.801	0.100	0.036	1.801	0.1
tCO2/t ($C_{2=} + C_{3=}$)			0.193			0.193
Total Emissions			1.116			0.309
tCO2/t ($C_{2=} + C_{3=}$)			2.159			0.598
Additional cost @ $35/t			$75.55			$20.92
Production cost ($/t)			$726.27			$869.19
Including carbon cost ($/t)			$801.83			$890.11

For the OPEN system all of the fuel used is assumed to be fuel oil. Including an operating allowance, this produces about 1 million tonnes of carbon dioxide for the production of 0.5 million tonnes of ethylene.

On a unit basis this is approximately 2 tonnes of carbon dioxide per tonne of olefins (as ethylene plus propylene). Operational losses add a further 100,000 tonnes of carbon dioxide making the total emissions of carbon dioxide per tonne of olefin an estimated 2.159 t/t.

If the cost of carbon emission is $35/tonne[6], then the additional cost is $75.55 million which lifts the ethylene production cost from $726/t to $801/t.

For the CLOSED system many of the by-products are recycled to either fuel or feedstock. The fuel in the furnaces is now hydrogen, methane and some fuel oil. A major portion of the fuel is hydrogen which does not appear in Table 7.6 because it does not contribute to carbon emissions.

After allowance is made for ethane flaring and the carbon dioxide emissions, the CLOSED system produces approximately 0.6 tonne of carbon dioxide per tonne of olefin. Using the same cost of carbon dioxide the result is that production cost is lifted from $869 to $890/tonne.

This use of hydrogen as a furnace fuel can dramatically change the relative economics of the OPEN and CLOSED systems as the cost of carbon emissions increases. This is demonstrated in Figure 7.10.

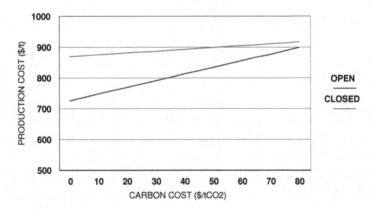

Figure 7.10: Ethylene from ethane - sensitivity to carbon emission cost

As is demonstrated by this figure, the use of hydrogen as a fuel can close the relative gap between the OPEN and CLOSED system, but this requires a carbon cost in excess of $80/tonne when the ethylene production cost is forced over $900/tonne.

---

[1] "Petrochemical Processes 2005", *Hydrocarbon Processing,* CD ROM, p. 71-81

[2] See Hydrocarbon Processes 2005, *Hydrocarbon Processes,* CD ROM, p. 71-75

[3] Muse Stancil & Co ethylene margins for US Gulf operations are published monthly in the *Oil & Gas Journal* and described in *Oil & Gas Journal*, September 16, 2002

[4] S. K. Kapur, A. S. Laghate and W. M. Nouwen, *Hydrocarbon Asia*, Sept. 1996, p. 110 discuses the use of the SPYRO model in optimising cracker operations.

[5] J. H. Vautrain, K. A. Barrow, *Oil & Gas Journal*, Sep. 6, 2004, p. 52; see also A. Keller, *Oil & Gas Journal*, Aug. 20, 2001, p. 75

[6] At the time of writing this about double the traded emissions cost for carbon dioxide but much less than the cost of disposal by geo-sequestration.

CHAPTER 8

LPG CRACKING

This chapter deals with the technology and economics of pyrolysis cracking of the natural gas liquids propane, butane and isobutane commonly known as LPG (liquefied petroleum gas). Often, LPG is separated into its main components and the cracking of these components will be discussed separately. Sometimes mixtures of the components are cracked and the outcome is the sum of the relative amounts of the individual components. In the 1970s and 80s, LPG was obtainable at a lower cost relative to naphtha and in that era many steam crackers were built to take advantage of this by cracking both naphtha and LPG. In some parts of the world where LPG feedstock remains relatively cheap it is still used in large volumes as a cracker feedstock. In other parts of the world "multi-feedstock" crackers can take advantage of periodic low cost supplies of LPG.

Basic Chemistry of LPG Cracking

The steam pyrolysis of LPG follows the same pathway of that for ethane, namely by a complex branching chain free radical mechanism. This can be divided into initiation, chain propagation and chain termination reactions. This gives rise to a large number of intermediates and products. As with ethane, products of higher carbon number than the feed are formed.

Because the cracking process involves the rupture of carbon-carbon bonds, products of the same carbon number as the feedstock are low in concentration. Thus for propane, the major product of cracking is ethylene and methane rather than propylene. Normal butane gives more propylene, but the main end product is again ethylene.

145

The branched nature of isobutane, however, makes propylene the major product. Table 8.1 gives the single pass yields for ethane, propane, normal and isobutane to illustrate these points.

Table 8.1: Products from the Cracking of Gaseous Feeds

FEED	ETHANE	PROPANE	n-BUTANE	ISOBUTANE
Hydrogen	3.72	1.56	1.49	1.08
Methane	3.47	23.65	19.9	16.56
Acetylene	0.42	0.77	1.07	0.72
Ethylene	48.82	41.42	40.59	5.65
Ethane	40	3.48	3.82	0.88
allene/propyne	0.2	1.09	1.07	2.34
Propylene	0.99	12.88	13.64	26.35
Propane	0.03	7	0.48	0.38
Butadiene	1.33	2.82	4.13	1.49
Isobutene				19.6
n-butenes	0.25	0.89	1.92	
Isobutane				20
n-butane			4	
py-gasoline	0.46	1.37	3.24	2.35
BTX	0.31	3.07	5.25	2.4
TOTAL	100	100	100.6	99.8

The practical outcome is re-emphasised in Figure 8.1, which illustrates the relative amounts of ethylene to propylene when cracking LPG feedstock to that of the liquid feeds such as naphtha and vacuum gas oil (VGO).

Practical Aspects of LPG Cracking

Because the formation of heavy liquid products is low, LPG is cracked in a very similar process to ethane cracking. Often LPG can be co-fed to the pyrolysis furnace with ethane and there is no need for an additional process plant.

However, if the LPG is from a refinery operation or downstream petrochemical production, olefins are often present and these can lead to increased fouling of the furnace. To prevent this, LPG streams containing olefins are often hydrogenated prior to cracking.

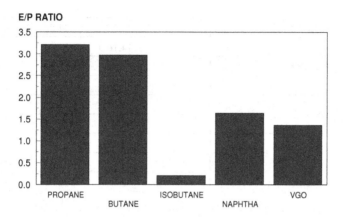

Figure 8.1: Ethylene/propylene ratio for various feedstocks

Economics of Propane Cracking

Although the cracking of propane is similar to ethane cracking producing a similar product slate, crackers designed to crack only propane cost more due to the larger size of plant handling the heavier product and the inclusion of equipment to upgrade the propane stream.

Following the same methodology for the cracking of ethane, the production cost of ethylene by propane cracking in an OPEN system is shown in Table 8.2. In this scenario, all of the products are on-sold to downstream operations or valued at an opportunity cost.

Table 8.2: Economics of Propane Cracking

	kt/y	PJ/y		
PROPANE PRICE			$/GJ	11.17
			$/tonne	561.9
	kt/y	PJ/y	MM$/y	MM$/y
CAPEX			$832.75	
OPEX (10% CAPEX)			$83.27	
RETURN ON WORKING CAP. (10%)			$9.12	
RECOVERY (10%DCF, 20y, FACTOR 0.143)			$119.08	

Table 8.2 (continued)

FEEDSTOCK & FUEL PURCHASES				
Feedstock	1129	56.79	$634.39	$634.39
Operating Feed (5.5%)	62	3.12	$34.89	$34.89
Fuel	330	14.16	$118.04	$118.04
Operating Fuel (3%)	10	0.42	$3.54	$3.54
TOTAL	1531	74.49	$790.85	$790.85
OUTPUTS				
Ethylene	500	25.15		$639.00
Propylene	155	7.58		$188.17
BTD/C4 olefins	45	2.16	$26.91	$26.91
Gasoline	17	0.79	$10.99	$10.99
Hydrogen	19	2.69	$43.11	$43.11
				$0.00
Methane	295	16.37	$130.98	$130.98
Propane	98	4.93	$55.07	$55.07
Butane	0	0	$0.00	$0.00
Fuel Oil	0	0	$0.00	$0.00
TOTAL	1129	59.68	$267.05	$1,094.22
THERMAL EFFICIENCY (%)		80.1%		
ANNUAL COSTS			$1,002.33	$883.25
BYPRODUCT CREDITS			$267.05	
UNIT ETHYLENE PRODUCTION COST ($/t)			$1,136.03	
MARGIN				$210.97
			$/t	$322.10
			c/lb	14.61
THERMAL EFFICIENCY		81.1%		
ANNUAL COSTS			$987.64	$850.67
BYPRODUCT CREDITS			$263.35	
UNIT ETHYLENE PRODUCTION COST ($/t)			$1,098.10	
MARGIN				$255.64
			$/t	$382.69

The first row of the table gives the price of propane which is linked to the prevailing crude oil price which is assumed to be $70/bbl. By-products are also priced with this value of crude oil. The scale is set for the production of 500,000 t/y of ethylene with other outputs given in the first column. The other main products are propylene (155 kt/y), a butadiene and C4 stream (45 kt/y), pyrolysis gasoline (17 kt/y) and hydrogen (19 kt/y). A large portion of the overall cracking stoichiometry for propane cracking is to produce ethylene and methane, rather than ethylene and hydrogen for ethane cracking. The consequence is that propane cracking produces much less hydrogen (19 kt/y versus 66 kt/y) but considerably more methane (295 kt/y) than ethane feedstock (40 kt/y).

The second column gives the calculation of the thermal efficiency of the cracking operation which is 80.1%. This is higher than that for ethane cracking which reflects the overall lower cracking temperature. The third column gives the estimation of the ethylene production cost and the forth column an estimate of the operating margin with the price for ethylene at $1278/t and propylene at $1214/t.

Following the basic methodology, the production cost for ethylene is estimated at about $1136/tonne. Using the basic values of the products, the unit operating margin is estimated at $322/tonne (14.61 cents/lb). The latter calculation takes no account of non feed operating costs and capital charges. This result is inferior to the ethane cracking outcome principally due to the higher feedstock costs. As illustrated in Figure 8.2, the production costs are dominated by the cost of feedstock.

The representative data for the OPEN, CLOSED and for the integration with polymer production is illustrated in Table 8.3. This illustrates that for the closed system when some of the valuable by-products are degraded into fuel value streams, results in the operating margin are significantly reduced. This is restored to higher values when production is integrated with the downstream production of polymer as the product is sold.

Under these conditions, the cracking of propane is seen as a less profitable operation relative to other feedstocks. This agrees with the trend over the past decade that as the relative cost of propane has risen,

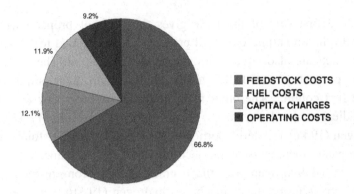

Figure 8.2: Production cost breakdown for propane cracking - OPEN system

Table 8.3: Comparison of Margins for Propane Cracking

	OPEN	CLOSED	polymer
UNIT ETHYLENE PRODUCTION COST ($/t)	$1,136.03	$1,218	$1,341
UNIT OPERATING MARGIN ($/t)	$322.10	$241.02	$603.94
UNIT OPERATING MARGIN (cents/lb)	14.61	$10.93	$27.39

there has been progressively less cracking of propane. However in
several parts of the world, propane is available at prices well below the
world parity price.

Sensitivity to Oil Price

As discussed previously, the world traded price for propane is
dependent on the prevailing price of crude oil, as are the values of
downstream products such as pyrolysis gasoline. Keeping all other
variables constant, the impact of changing the price of oil on the
economics of cracking propane is illustrated in Figure 8.3, which shows
how the production cost rises with increasing oil price.

As is illustrated by this figure, the cost of production for the three
scenarios considered is very sensitive to oil price. In this analysis there is
little difference between the three scenarios. On close inspection it is
seen that at low oil price, the OPEN system produces a lower cost as
minor by-products are degraded to fuel value in the CLOSED system.

However, as oil price rises the gap closes as increasing oil price impacts on the higher fuel cost in the OPEN system.

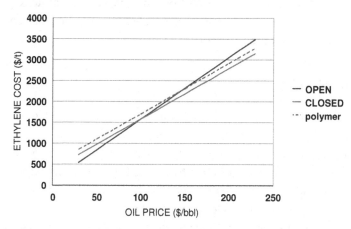

Figure 8.3: Sensitivity of ethylene production cost by propane cracking to oil price

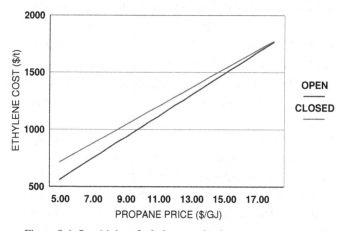

Figure 8.4: Sensitivity of ethylene production cost to propane price

Sensitivity to the Price of Propane

Keeping all other values the same, with oil at $70/bbl, the sensitivity of the production cost to the price of propane is illustrated in Figure 8.4. This plots the cost of propane from $1/GJ to $17/GJ approximately from $50 to $900/tonne.

The higher values correspond to the situation of high oil price and import parity pricing for propane. The lower end expresses the situation where propane is priced on a cost of recovery basis from large gas plants in the Middle East. These figures clearly indicate the competitive advantage of the latter operations.

Economics of Butane Cracking

The cracking of normal butane is very similar to that for propane and is accomplished in plants similar in design to that for ethane cracking. Capital costs are slightly higher (at $850 million for 500,000 t/y ethylene capacity, to account for the handling of heavier products and any feed hydrogenation that may be required. The basic statistics are shown in Table 8.4 for the OPEN system.

Relative to other gaseous feeds, butane cracking produces more of the heavier products; for instance in producing 500 kt ethylene, butane cracking produces more pyrolysis gasoline (40 kt) compared to propane (17 kt) and ethane (11 kt). Propylene is also produced in increased quantities: 168 kt versus 155 kt for propane. The temperatures required to produce the products are lower than that required for ethane, which results in the thermal efficiency rising to about 81% from 78% for ethane. Relative to propane, butane cracking has a slightly improved thermal efficiency.

Compared to propane cracking, despite a slightly higher capital cost, the ethylene production cost is lower at $1098/t versus 1136/t. The improvement is a consequence of a higher propylene and gasoline production and less methane. Together these improve the amount of by-product credit. Under the base case conditions with butane linked to the prevailing oil price set at $70/bbl, the operating margin is about $380/t.

With butane priced a world parity and oil at $70/bbl, feedstock prices dominate the production cost. This is illustrated in Figure 8.5 for an OPEN system (all products sold at opportunity value).

The sensitivity on the price of oil is illustrated in Figure 8.6 which plots the production cost for the OPEN and CLOSED system against the oil price.

Table 8.4: Statistics for n-Butane Cracking (OPEN)

BUTANE PRICE			$/GJ	10.89
			$/tonne	533.4
CAPEX			MM$	$850.79
OPEX (10% CAPEX)			MM$/y	$85.08
RETURN ON WOR. CAP			MM$/y	9.22
RECOVERY (10%DCF, 20y, FACTOR 0.143)			MM$/y	$136.98
FEEDSTOCK & FUEL PURCHASES				
	kt/y	PJ/y	MM$/y	MM$/y
Feedstock	1128	55.27	$601.70	$601.70
Operating Feed (5.5%)	62	3.04	$33.09	$33.09
Fuel	330	14.16	$118.04	$118.04
Operating Fuel (3%)	10	0.42	$3.54	$3.54
TOTAL	1530	72.89	$756.37	$756.37
OUTPUTS				
Ethylene	500	25.15		$639.00
Propylene	168	8.22		$203.95
BTD/C4 olefins	75	3.61	$44.85	$44.85
Gasoline	40	1.86	$25.85	$25.85
Hydrogen	18	2.55	$40.84	$40.84
Methane	259	14.37	$115.00	$115.00
Propane	19	0.96	$10.68	$10.68
Butane	49	2.4	$26.14	$26.14
Fuel Oil	0	0	$0.00	$0.00
TOTAL	1128	59.11	$263.35	$1,106.30

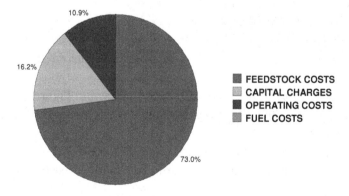

Figure 8.5: Production cost breakdown for butane cracking - OPEN system

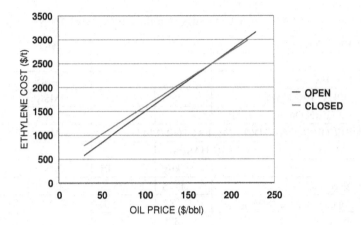

Figure 8.6: Sensitivity of ethylene production cost by butane cracking to oil price

Table 8.5: Comparison of Margins for Butane Cracking

	OPEN	CLOSED	POLYMER
CAPEX (MM$)	850	850	1,084
UNIT PRODUCTION COST ($/t)	1,098	1,246	1,353
UNIT OPERATING MARGIN ($/t)	382	236	605
UNIT OPERATING MARGIN (cents/lb)	17.4	10.71	27.40

At low oil price, the OPEN system produces a lower cost as minor by-products are degraded to fuel value in the CLOSED system. However, as oil price rises the gap closes as increasing oil price impacts on the higher fuel cost in the OPEN system.

The POLYMER case is developed from the closed system when the ethylene and propylene streams are integrated with polymer production. The relevant statistics are given in Table 8.5 comparing all three scenarios.

The impact of the price of butane is shown in Figures 8.7, which shows how the ethylene production cost are influenced by a change in the butane price.

The graph illustrates the lowering of the production cost as the butane price falls away from the parity price influenced by the price of oil.

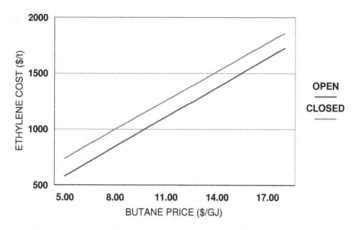

Figure 8.7: Sensitivity of ethylene production cost to butane price

Like propane, the use of butane as a cracker feedstock has fallen as the world price has become more influenced by the rise in oil price. However, in many parts of the world refinery operations are adopting new fuel standards that discourage the use of refinery butane in gasoline blending. This may make available larger volumes of butane for petrochemical cracking. Since there may be no alternative (full priced) use, butane may become available at a discount to the prevailing world parity price. As the figures illustrate, this could be a profitable operation for appropriately sited petrochemical plants.

Isobutane Cracking

Natural gas butane is usually dominated by the straight chain or normal isomer. There is some isobutane present and in some fields, the concentration can be high, but rarely the dominant isomer. However, refinery catalytic cracking produces butane streams where the main isomer is branched, namely isobutane. The concentration is often 90% of the stream.

Isobutane is a valuable refinery intermediate for the production of alkylate gasoline and the demand for alkylate is set to rise as it is an attractive blend component for gasoline made to the new fuel standards now being introduced. Where the demand is large enough, normal butane can be isomerised into the branched isomer to capture its generally higher value.

Because the cracking product slate is quite different between normal and isobutane, it is important to take cognisance of the isobutane concentration in a butane stream used for cracking. The difference is illustrated in Table 8.6.

Table 8.6: Product Slate for Butane Isomers

	normal	Iso
Ethylene	41.66	7.67
Propylene	14.71	34.57
BTD/C4 olefins	6.05	25.41
Gasoline	8.49	5.72
Hydrogen	1.49	1.30
Fuel Gas	23.72	21.01
Propane	0.48	0.46
Butane	4.00	3.86
Fuel Oil	0.00	0.00
TOTAL	100.6	100

The data has been adapted from Table 8.1 to show the relative amounts of products at similar conversion levels. This indicates that the differences are mainly the considerably lower ethylene yield and higher propylene and C_4 olefin yield (with the branched isomer dominant).

Thus, an operation able to crack butane would be able to lift propylene yields if the increased amounts of isobutane were to be fed into the system. This could be by either isobutane purchases or isomerisation of n-butane feedstock. This gives a gas cracking operation some flexibility in altering the ethylene/propylene split which is otherwise difficult with gaseous feedstock.

Carbon Dioxide Emissions in LPG Cracking

In the CLOSED scenario (in which some of the products are used as fuel or feedstock) there is an excess of methane and hydrogen over that required to fuel the cracker furnaces, including the operating allowance. This is the case for both propane and butane. This excess is due to the increased production of methane. For ease of analysis, it is assumed that the fuel gas in excess of requirements is flared. This maintains the assumption of the CLOSED system that there is no market for excess hydrogen or methane available.

The analysis follows the method as described previously for ethane cracking and the results are illustrated in Figure 8.8.

Unlike the ethane case where the use of hydrogen as furnace fuel greatly reduces the carbon emissions in the CLOSED system, the lower hydrogen production and increased production of methane results in only a small lowering of emission intensity relative to the OPEN system.

The graph illustrates the sensitivity to a carbon dioxide emission cost, which for a cost of \$35/t typically translates to an increase of about \$50/t of olefin product.

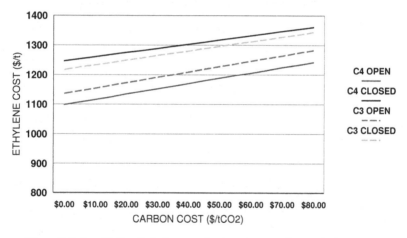

Figure 8.8: Sensitivity of ethylene production cost to carbon emission cost

CHAPTER 9

LIQUID FEEDSTOCK CRACKING

This chapter focuses on the economics of cracking naphtha and gas oil. The cracking of liquid feedstock produces most of the world's ethylene. This is dominated by naphtha cracking, the character of which has been discussed previously. Where available, there is some cracking of gas oil.

The cracking of naphtha is carried out in all regions. The nominal capacity of the operations ranges from about 250 kt/y ethylene to operations producing over 1 million tonnes ethylene. There is good economy of scale and today's world scale crackers have a typical scale of 500 to 1000 kt/y ethylene, typically 850 kt/y.

A distinguishing feature of naphtha operations relative to gas feedstock crackers is the production of heavier products. This requires a considerable amount of process plant at the front end and the back end of the plant to handle them. A typical process block flow is illustrated in Figure 9.1.

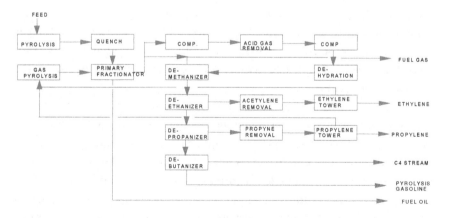

Figure 9.1: Typical unit operations for pyrolysis of naphtha and gas oil

Feedstock (after pre-treatment if necessary) is passed along with steam to the pyrolysis furnace. This cracks the compounds in the naphtha, producing a full range of products which are extremely complex. As with gas feedstock, heavier products are produced, but in increased volumes. After quenching a primary fractionator (not present in gas crackers) separates the heavy pyrolysis fuel oil from the cracked gases.

One of the issues that concern liquid feedstock cracking operations is a higher rate of fouling. This is not only a consequence of heavier coke forming precursors, but also as a consequence of long lived free radicals which act as agents for the formation of a polymer (often referred to as pop-corn polymer) in the primary fractionator and downstream units. For instance, free radicals based on styrene or indene have sufficiently long half-lives to pass from the pyrolysis section into the primary fractionator. These can concentrate in this unit and produce polymer (free radical polymerisation) when sufficient amounts of suitable olefins are present, in particular styrene itself and di-olefins such as cyclo-pentadiene or butadiene.

After primary separation, the cracked gases are compressed, acid gases are removed, the product gases are recompressed and dried then light gases – methane and hydrogen – are removed in the de-methaniser tower. These light fuel gases are often separated in a cryogenic cold-box. This is adversely affected if mercury is present in the naphtha; there has been a move in recent times to introduce mercury removal as a pre-treatment step to the naphtha feed[1]. This is usually not an issue for gas feedstock, because large gas plants separating ethane often employ cold-box technology and hence require mercury removal upstream of the petrochemical plant.

This is followed by separation of the cracked gases according to carbon number in much the same was as that for gaseous feedstock. As well as larger volumes in of C_3 products, there are large volumes of material in the C_4 and pyrolysis gasoline fractions. These are sent to downstream processing units for further separation and upgrading.

One of the features of liquid feedstock cracking is the production of large volumes of ethane. These are collected by the de-ethanizer tower and sent back to the pyrolysis section where a separate furnace operating at higher temperature is dedicated to ethane cracking. This ultimately cracks all of the ethane produced to extinction.

Naphtha cracking is nowadays rarely stand alone and is most often integrated into downstream operations. Naphtha crackers are often the central features of the world's large integrated chemical complexes. Figure 9.2 illustrates a typical complex which produces a wide variety of chemicals sold to downstream producers of consumer goods[2].

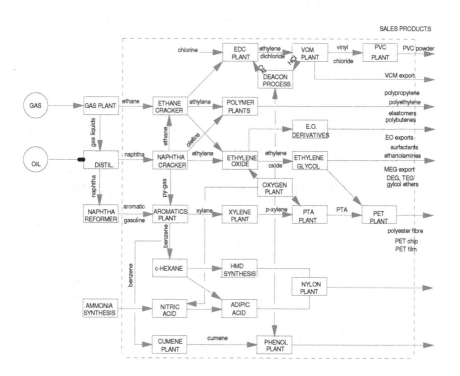

Figure 9.2: Typical large integrated petrochemical facility

Economics of Naphtha Cracking

Unlike gas feedstock cracking the economics cracking of liquid feedstock such as naphtha is complicated by the ability to change the product slate by changing the cracking conditions (severity) and to select naphtha of differing compositions. First the basic issues are discussed which explore the underlying economic impacts.

Base Case for Naphtha Cracking

The base case is developed for the production of 1 million tonnes of ethylene per year, which is typical of the scale for naphtha-crackers being established in 2008. The base case statistics (oil at $70/bbl) for the OPEN system (all products sold) are given in Table 9.1.

The capital for the basic operation producing 1000kt/y ethylene is about $1,700 million. This is for a stand-alone plant producing the products as detailed and does not, for instance, include downstream plant for separating the C4 olefin stream.

As well as 1000 kt/y, the cracking produces about 500kt/y propylene, 254kt/y of mixed C4 olefins, 750kt/y pyrolysis gasoline and 46kt/y hydrogen. The cracking also produces 480kt/y methane, 126kt/y LPG and 148kt/y of pyrolysis fuel oil. The sale of these by-products considerably improves the process economics by facilitating economies of scale in downstream process plants and by boosting by-product credits to over $1,000 million/year.

With oil at the $70/bbl mark, the estimated production cost is $1,209/tonne with an operating margin of about $229/tonne.

The breakdown of the production costs is illustrated in Figure 9.3, which illustrates the dominance of feedstock cost in the breakdown of the production cost which represents over 72% of the costs for a green-fields stand alone plant.

A CLOSED operation can be modelled in which all products other than ethylene, propylene and pyrolysis gasoline are recycled. The propane and C_4 streams are recycled to the feed side of the pyrolysis furnace and the hydrogen, methane and fuel oil to the fuel side of the furnace. This offsets some of the need for fuel but reduces the by-product credits to $482 million/year and this raises the production cost to $1227/tonne and reduces the operating margin to $211/tonne as valuable by-products are reduced to fuel value. The comparative cost breakdown is illustrated in Figure 9.4.

Table 9.1: Statistics for Naphtha Cracking

CAPEX		MM$	$1,708.80	
OPEX (10% CAPEX)		MM$/y	$170.88	$170.88
RETURN ON WOR. CAP		MM$/y	$24.77	$24.77
RECOVERY (10%DCF, 20y, FACTOR 0.143)		MM$/y	$244.36	
	kt/y	PJ/y	MM$/y	MM$/y
FEEDSTOCK & FUEL PURCHASES				
Feedstock	3312	159.31	$1,980.40	$1,980.40
Operating Feed (5.5%)	182	8.76	$108.92	$108.92
Fuel	926	39.73	$331.21	$331.21
Operating Fuel (3%)	28	1.19	$9.94	$9.94
TOTAL	4448	208.99	$2,430.47	$2,430.47
OUTPUTS				
Ethylene	1000	50.30		$1,278.00
Propylene	512	25.04		$621.57
BTD/C4 olefins	254	12.22	$151.88	$151.88
Gasoline	746	34.61	$482.19	$482.19
Hydrogen	46	6.52	$104.36	$104.36
Methane	480	26.64	$213.12	$213.12
Propane	40	2.01	$22.48	$22.48
Butane	86	4.21	$45.87	$45.87
Fuel Oil	148	6.35	$52.94	$52.94
TOTAL	3312	167.91	$1,072.84	$1,072.84
THERMAL EFFICIENCY (%)		80.34%		
ANNUAL COSTS			$2,870.48	$2,626.12
BYPRODUCT CREDITS			$1,072.84	
UNIT ETHYLENE PRODUCTION COST ($/t)			$1,209.42	
MARGIN				$346.29
			$/t	$229.03
			c/lb	10.39

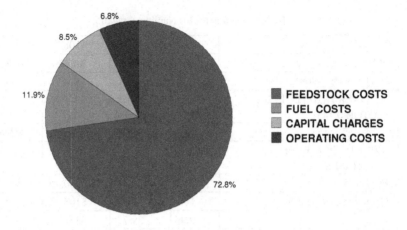

Figure 9.3: Breakdown of ethylene production cost using naphtha
feedstock – OPEN system

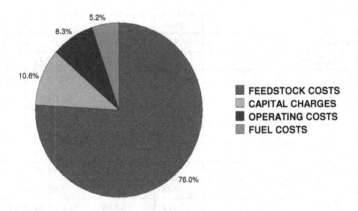

Figure 9.4: Breakdown of ethylene production cost using naphtha
feedstock – CLOSED system

Sensitivity to Oil Price

The price of oil influences the cracking economics by changing
both the price of feedstock and fuel and changing the value for pyrolysis
gasoline and fuel oil. The sensitivity of the unit production is illustrated
in Figure 9.5.

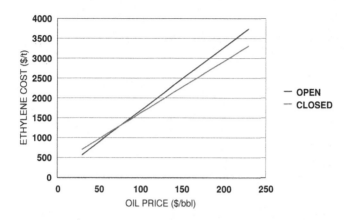

Figure 9.5: Sensitivity of ethylene production cost from naphtha to oil price

With oil at $70/bbl, the production cost is about $1,200/t for both OPEN and CLOSED scenarios. This sensitivity analysis suggests that as the oil price rises above $70/bbl the CLOSED system becomes progressively more economic (representing the upgrading of some of the by-products and naphtha to gasoline). However, this may not eventuate in practice because this modelling may be over-simplified with recycled LPG being modelled as giving the same product slate as naphtha.

Sensitivity to the Relative Value of Ethylene and Propylene

One of the main features of cracking liquids is the high yield of propylene. In the world market there are often marked regional differences in the relative demand and hence price of ethylene and propylene. In the EU, propylene generally trades at a lower value that ethylene, whereas in the Far East this is reversed.

The basic economics are quite sensitive to the relative value of propylene to ethylene. This is illustrated in Figure 9.6, which shows how the production cost of ethylene changes with the relative value propylene and ethylene.

The base case economics assumes the unit values of ethylene and propylene have a ratio of 0.95, similar to typical EU experience. The graph shows that when the propylene value is higher than the value of ethylene there is a progressive fall in the production cost of ethylene.

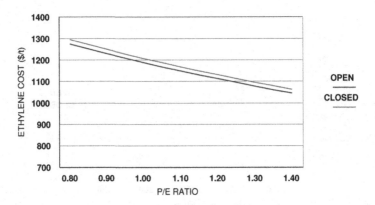

Figure 9.6: Sensitivity of ethylene production cost from naphtha to P/E price ratio

Integration with Downstream

The above economic analysis indicates that stand-alone naphtha cracking operations have a high unit production cost. Nowadays, the vast majority of crackers operate on an integrated basis producing downstream saleable goods, or intermediates, rather than olefins. This is not only polymers and resins (polyethylene and polypropylene), but because of the large amount of pyrolysis gasoline, downstream processing to produce aromatics – benzene, toluene and xylene (BTX) are very common. In large integrated complexes, the latter are further processed to styrene, nylon and polyester.

The statistics for a naphtha cracker integrated with polymer and BTX production are illustrated in Table 9.2. The complex is based on the CLOSED case and only ethylene, propylene and pyrolysis gasoline pass to downstream processing. For brevity it is assumed the polymers are produced at 100% yield and require no other feedstock. The pyrolysis gasoline is passed to an aromatics extraction plant and produces 298 kt/y benzene, 149 kt/y toluene and 52 kt/y xylene. The rest of the pyrolysis gasoline (246 kt/y) produces a raffinate, which is sold as a gasoline.

This integrated plant will cost about 2,500 million compared to the $1,708 million of the cracker only option. However, with the by-products valued with oil at $70/bbl or as those typical for 2007, the polymer production costs are about $1,260/tonne. This is about the same product

Table 9.2: Naphtha Cracking to Polymer and BTX

Capital Costs	kt/y	MM$	MM$/y
Naphtha cracker Capex		1708.80	
Polyethylene Capex	1000	466.76	
Polypropylene Capex	512	238.98	
BTX Capex	746	84.21	
Total Capex		$2,498.75	
Opex (10% Capex)			$249.88
Return on Working Capital			26.32
ROC (DCF, 20y, FACTOR 0.143)			$357.32
Feedstock and Fuel Purchases	kt/y	MM$/y	MM$/y
Feedstock	2929	$1,751.12	$1,751.12
Operating Feed (5.5%)	182	$108.92	$108.92
Fuel	5	$1.78	$1.78
Operating Fuel (3%)	28	$9.94	$9.94
Total	3143	$1,871.76	
Polyethylene	1000		$1,702.40
Polypropylene	512		$814.28
Gasoline	246	$159.12	$159.12
Benzene	298	$313.62	$313.62
Toluene	149	$121.75	$121.75
Xylene	52	$46.84	$46.84
Total	2258	$641.33	$3,158.01
Annual Costs		$2,505.27	$2,147.95
By-product Credits		$641.33	
Polymer Production Cost ($/t)		$1,260.86	
MARGIN			$1,010.06
		$/t	$668.03

production cost as the CLOSED case for a more valuable product. With polyethylene prices at the $1,700/tonne mark, the operating margin is $668/tonne considerably higher that the $220/tonne for a stand alone cracker producing only olefins. The breakdown of the production costs are illustrated in Figure 9.7.

This illustrates that the further integration with downstream operations results in a lessening of the dependence of feedstock compared to the non integrated cases above. These results clearly indicate the benefits of integration with downstream. Further improvements in profitability can be made by producing higher-grade polymers such as LLDPE and further upgrading of the raffinate from the aromatics plant. The sensitivity of this integrated case to the prevailing oil price is illustrated in Figure 9.8.

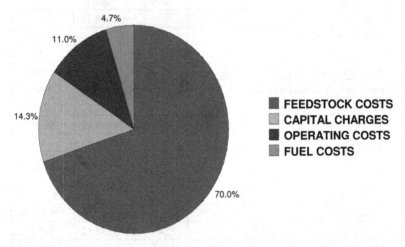

Figure 9.7: Production cost breakdown for polymer and BTX using naphtha

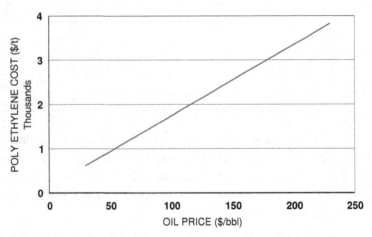

Figure 9.8: Sensitivity of polymer production cost from naphtha to oil price

Impact of Cracking Severity

A distinguishing feature of liquid feedstock cracking in contrast to cracking using a gas feedstock is the ability to increase the ethylene to propylene ratio in the product slate by increasing the cracking temperature. This is generally known as increasing the cracking severity. The outcome is illustrated in Table 9.3, which compares the same naphtha feedstock cracked at high and low severity to produce 500 kt/y ethylene.

Table 9.3: Naphtha Cracking at High and Low Severity

	HIGH SEVERITY		LOW SEVERITY	
	kt/y	PJ/y	kt/y	PJ/y
FEEDSTOCK & FUEL PURCHASES				
Feedstock	1613	70.97	1872	82.37
Operating Feed (5.5%)	98	3.91	103	4.54
Fuel	688	29.52	611	26.21
Operating Fuel (3%)	21	0.89	18	0.79
TOTAL	2410	105.28	2604	113.91
OUTPUTS				
Ethylene	500	25.15	500	25.15
Propylene	193	9.44	310	15.16
BTD/C4 olefins	105	5.05	174	8.37
Gasoline	351	16.29	523	24.27
Hydrogen	26	3.69	21	2.98
Methane	300	16.65	200	11.1
Propane	20	1.01	20	1.01
Butane	22	1.08	66	3.23
Fuel Oil	96	4.12	58	2.49
TOTAL	1613	82.46	1872	93.76
THERMAL EFFICIENCY (%)		78.32%		82.31%

The high severity option uses a high temperature to produce 500 kt/y ethylene. This requires more fuel but less naphtha feedstock than the low severity case. This is reflected in the lower thermal efficiency for higher severity products.

The high severity operation increases the production of ethylene by increasing the cracking of heavier components and by-products. Consequently all of the other products, other than methane and hydrogen, are reduced relative to the low severity operation.

These changes result in lower feedstock costs, but higher fuel costs and lower by-product credits for the lower severity case. The typical range in outcomes as the oil price changes for the OPEN system is illustrated in Figure 9.9. This indicates that with oil at $70/bbl and using factored base case statistics for capital, feed and by-product credits, the difference in ethylene production costs range over $150/t. Figure 9.10 illustrates the effect of changing the value of propylene relative to ethylene (P/E) on the ethylene production costs. These figures illustrate the advantage of low severity operation in producing high valued propylene with the concomitant advantage of producing higher volumes of pyrolysis gasoline.

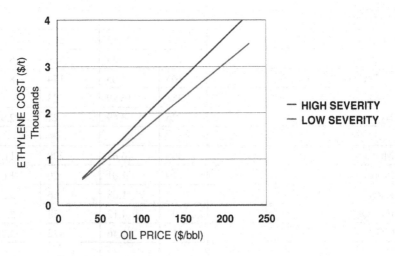

Figure 9.9: Sensitivity of ethylene production cost from naphtha to oil price – OPEN system

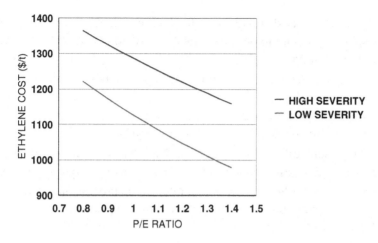

Figure 9.10: Sensitivity of ethylene production cost from naphtha to P/E
price – OPEN system

Gas Oil Cracking

In some parts of the world gas oil is cracked to produce
petrochemicals. Gas oil and heavier feedstock is produced from the
heavier end of the crude oil barrel, as illustrated in Figure 9.11.

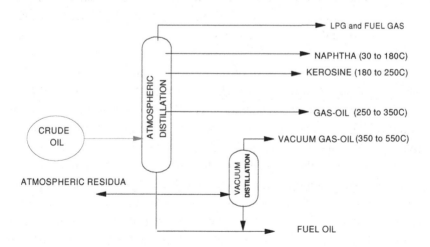

Figure 9.11: Liquid feedstocks from crude oil

Crude oil is distilled in a distillation column operating near to atmospheric pressure to produce naphtha (b.p. 30°C – 180°C) and gas oil (b.p. 25°C – 360°C). The bottoms of the column, known as atmospheric residua, are passed to a vacuum distillation column which produces vacuum gas oil (b.p. 350°C – 550°C). Any of these distilled liquid feeds can be used to produce petrochemicals.

If gas oil and heavier feedstock are used as a cracker feedstock, the most important difference relative to naphtha is in the production of heavier materials, especially pyrolysis fuel oil, which requires more plant and equipment for handling. Fouling rates in the process plant exposed to the heavier materials are higher than those experienced for cracking naphtha.

However, gas oil is used in refineries to produce diesel transport fuel and has consequently a high value and a product slate too poor to make it an attractive feedstock relative to naphtha. In recent years, many gas oil crackers have been reconfigured to crack lighter feedstock or heavier feedstock such as atmospheric residual fuel (b.p. > 360°C).

The enhanced fouling rates and metal contamination (from the crude oil) generally makes atmospheric residua unsuitable as a cracker feedstock. However, some crude oils produce a waxy residual of low metal content (often referred to as low sulphur waxy residua, LSWR). Although more expensive than fuel oil, LSWR is considerably cheaper than gas oil and is an attractive feedstock for some gas oil cracker operations.

Gas oil cracking has all of the characteristics of naphtha cracking. The typical statistics are given in Table 9.4. Relative to naphtha, the gas oil cracking requires considerably more feed for the same ethylene output (500 kt/y) and at the same time produces increased volumes of pyrolysis gasoline and more particularly pyrolysis fuel oil. This requires an increase in the fixed capital to naphtha for the same scale of operation.

For any given oil price, gas oil sells at a higher value because of its demand for motor diesel, whereas fuel oil sells at a price lower than the crude oil price (WTI, Tapis, Brent etc.). LSWR is linked to fuel oil prices usually (but not always) and sells at a slightly higher price but still well below the crude oil price.

Table 9.4: Statistics for Gas Oil Cracking at High and Low Severity

	HIGH SEVERITY		LOW SEVERITY	
INPUTS	kt/y	PJ/y	kt/y	PJ/y
Feedstock	1864	82.02	2479	109.08
Operating Feed (5.5%)	103	4.51	136	6
Fuel	625	26.81	650	27.89
Operating Fuel (3%)	19	0.8	20	0.84
TOTAL	2610	114.14	3285	143.8
OUTPUTS				
Ethylene	500	25.15	500	25.15
Propylene	173	8.46	359	17.56
BTD/C4 olefins	100	4.81	197	9.48
Gasoline	396	18.37	531	24.64
Hydrogen	21	2.98	21	2.98
Methane	263	14.6	205	11.38
Propane	27	1.36	25	1.26
Butane	19	0.93	82	4.02
Fuel Oil	365	15.66	559	23.98
TOTAL	1864	92.32	2479	120.43
THERMAL EFFICIENCY (%)		80.88		83.75

The sensitivity of the production economics for LSWR is illustrated in Figure 9.12 for the open system (all products sold at market price). This shows the production cost is very sensitive to the prevailing price of the feedstock. However, despite an increase in capital relative to the naphtha case, the production cost for ethylene is lower than the naphtha case due to the marked lower feedstock price by about $200/t for each of the scenarios.

Again the production cost is sensitive to the relative price of propylene and ethylene which is illustrated in Figure 9.13. One of the features of heavy oil cracking is the high propylene yield at low severity. This serves to further reduce ethylene production cost as the relative value of propylene rises.

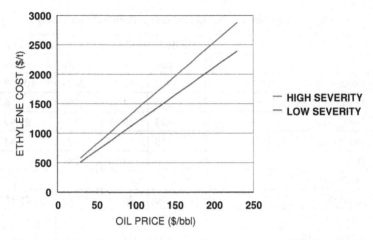

Figure 9.12: Sensitivity of ethylene production cost from LSWR to oil price

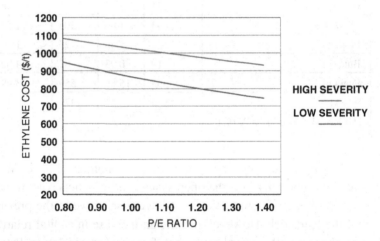

Figure 9.13: Sensitivity of ethylene production cost from LSWR to P/E price

The LSWR's availability is limited and can show high price volatility[3] so that few operations are totally dedicated to this feedstock. A more common scenario is to adapt some of the furnaces operations of a naphtha cracker to take heavier fuels on an occasional basis when the feedstock becomes available.

Carbon Emissions from Naphtha and LSWR Cracking

As well as OPEN and CLOSED operations, for liquid feedstocks there are several issues to be addressed which are not relevant or have little impact on the economics of olefin production from gas feedstocks. These are:

- How emissions are to be distributed across all of the products produced, and
- The impact of severity of operation on the emissions

OPEN versus CLOSED operations

Table 9.5 gives the statistics for cracking naphtha (1 million tonnes of ethylene/year) for the OPEN and CLOSED systems. The first column gives the mass of fuel and feed contributing to on-site emissions. The second column, the energy (HHV basis) for fuel and feed used. The third column gives the carbon dioxide emissions and when calculated on a unit basis this emission is attributed to the principal olefin products (ethylene and propylene). The fourth column gives the pertinent unit emissions when the carbon dioxide emission is distributed across all of the products.

The sensitivities to the carbon disposal cost are shown in Figure 9.14. Within error of this type of analysis, these cost curves are the same with the suggestion that as carbon disposal cost rises, the CLOSED system becomes more cost competitive than the OPEN system because high carbon intensity fuels are backed out in favour of hydrogen and methane.

Emission Distribution across Products

Also given in Table 9.5 is the effect of distributing the cost of carbon emissions across only the olefins (ethylene and propylene) versus distributing this cost over all of the saleable products, i.e. that pyrolysis gasoline and other products should receive some of the carbon charge. The cost curves for the various scenarios are shown in Figure 9.15.

Table 9.5: Carbon Emissions from Naphtha Cracking

	OPEN							CLOSED
	Mt/y	PJ/y	Mt CO2/y	Mt CO2/y	MT/y	PJ/y	Mt CO2/y	Mt CO2/y
Fuel Used			Olefins only	All products			Olefins only	All products
Natural gas					26.640	1.367		
LPG					0	0		
Fuel oil	0.954	40.917	2.852		7.754	0.512		
Total	0.954	40.917	2.852		34.394	1.878		
tCO2/t products			1.886	0.861			1.242	0.832
Flaring								
Naphtha	0.182	8.762	0.578		0.182	8.762	0.578	
tCO2/t products			0.382	0.175			0.382	0.256
Total Emissions			3.430				2.457	
tCO2/t product			2.269	1.036			1.625	1.088
			$/t	$/t				$/t
CO2 cost/t product			79.4	36.2			56.9	38.1
Ethylene production cost free of CO2 cost			1209.4	1209.4			1227.6	1227.6
Production cost including CO2			1288.8	1245.7			1284.5	1265.7

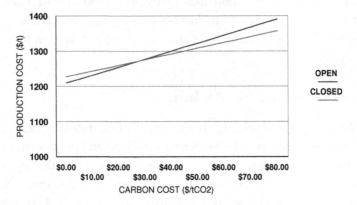

Figure 9.14: Sensitivity of ethylene production cost from naphtha to carbon emission cost – OPEN and CLOSED systems

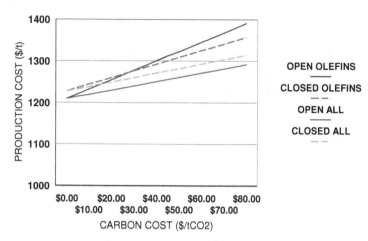

Figure 9.15: Sensitivity of ethylene production cost from naphtha to carbon emission cost distributed over olefins only and over all of the products

This figure illustrates the benefit of being able to distribute the cost of carbon emissions across all of the products rather than just the olefin stream.

How this would work in practice is moot since for the most part these by-products (pyrolysis gasoline, fuel oil, and methane) compete in markets where alternative sources are produced by other, potentially less carbon intense means. It may be that all of the additional cost of production cannot be passed on to the by-product stream and the impact on the olefin production cost will be somewhere between the two levels.

Severity

The impact of severity is illustrated in Table 9.6 which gives the four scenarios of interest. As can be seen, there is over 1 t/t of carbon to be saved by going from a high severity OPEN system to a low severity CLOSED system.

Table 9.6: Carbon Emissions at High and Low Severity

System	OPEN	OPEN	CLOSED	CLOSED
Severity	High	Low	High	Low
CO2 (t/t olefin)	2.762	2.029	2.053	1.520

---

[1] Axens, "Petrochemical Processes 2005", *Hydrocarbon Processing, ,*CD ROM, p. 82

[2] K. Weissermel, H.-J. Arpe, "Industrial Organic Chemistry" VCH Publishers, 2 edition 1993 details many downstream process operations.

[3] LSWR finds use in power generation in the Far East which is constrained by the sulphur emissions. It is used to displace higher sulphur fuel oils and coal in periods of high power demand.

CHAPTER 10

OTHER ROUTES TO OLEFINS FROM HYDROCARBONS

In this chapter we consider other commercially important routes to the production of olefins other than by thermal steam cracking. Alternative processes generally involve catalytic processes rather than homogeneous gas-phase cracking. Although there have been proposals to develop catalytic processes for the production of ethylene, most of these alternative processes aim to produce propylene rather than ethylene. Some process economics of some of these routes have been compared by Houdek and Anderson[1] and Nextant Inc.[2]

Fluid Cat-Cracking (FCC)

Fluid catalytic cracking, fluid cat-cracking or FCC, is a common oil refinery process. The duty of an FCC unit is to take a heavy low value gas oil or fuel oil and convert this to higher valued liquid products, particularly gasoline blend-stock. The process also produces diesel fuel blend-stock and a gas by-product stream. The gaseous by-products are rich in olefins and in particular propylene and isobutene. Ethylene is a minor component.

FCC unit operations are the central feature of many refineries. Often it is physically the largest unit present in a refinery. It has been under development over many decades and has developed to use residual fuel oil rather than gas oil as the feedstock. In another development, catalysts have been developed to lift the level of propylene produced so that an FCC unit can produce considerable quantities of propylene as well as gasoline blend-stock[3]. FCC units produce a major portion of the chemical propylene in many countries such as the USA and Australia.

Fluid Cat-Cracking Operations

Figure 10.1 shows the general layout of the process. The feedstock, usually residual fuel oil, is heated to about 500°C prior to entering the unit. The feed enters a pipe where it is mixed with hot catalyst from the regenerator. Feed undergoes immediate and rapid cracking as it rises in the pipe (known as the riser) and enters the separation unit.

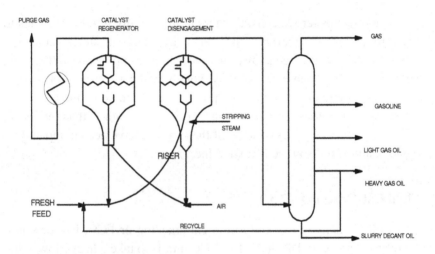

Figure 10.1: Schematic for a fluid cat-cracker

In the separator unit, steam separates the catalyst from the hydrocarbon products. The internals of the stripping unit contain cyclones which remove residual catalyst from the hydrocarbon fluids. The fluids flow to a distillation column which separates the products into various fractions.

The column bottoms (referred to as slurry decant oil) pass to heavy oil processing units or are blended into heavy fuel oil. The heavy gas oil fraction (sometimes called heavy cycle oil) is recycled or used as blend-stock for heavy diesel or industrial fuel oil. The light gas oil (sometimes called light cycle oil) is used for diesel blend-stock. The most important product is the gasoline component which goes to gasoline blending. The gas fraction (which can be rich in propylene) goes to the refinery gas plant where the olefins are separated.

Propylene exiting the gas plant is suitable for many refinery and chemical operations. For polypropylene manufacturing further purification is required to protect downstream units from traces of acetylenes and allene which may be produced.

Catalyst falling to the bottom of the separator is passed by an air-lift to a regenerator unit. Here, the air burns carbon deposited on the catalyst and reheats the catalyst before its return to the riser pipe. In some systems, and to prevent overheating in the regenerator, the oxidation of carbon proceeds only to carbon monoxide and carbon monoxide combustion to carbon dioxide occurs in a second regenerator.

Excessive heat generation in the regenerator is a particular problem when using residual feed when coke formation is higher. Residual fuel FCC operations generally have additional heat removal mechanisms in the regenerator. This can be steam raising coils or external catalyst coolers.

Gas expansion in the regenerator is used to produce power in a downstream gas turbine before the carbon dioxide is discharged to the atmosphere.

Catalysts

The basic catalysts used in FCC operations are highly acidic solids. Prior to FCC, cracking technology depended on amorphous silica-alumina which has an acidic character and upon which the chemistry is based. The discovery of crystalline alumina-silicates, the zeolites, that are stable at the cracking temperatures revolutionised cracking technology. The main catalyst is a stabilised zeolite-Y crystals which are held within an amorphous silica-alumina matrix. Stabilisation of zeolite-Y involves exchanging with rare earths (commonly noted as RE-Y) followed by a steaming process.

Zeolite-Y has very large diameter pores which permit the entry of the large feedstock molecules. These interact with the acid sites and at the temperature of operation (typically 550°C) result in cracking of the molecules into smaller molecules. As well as cracking, hydrogen transfer occurs so that the products produced are smaller molecules, with higher hydrogen to carbon ratio than the feedstock, and carbon, which is in the form of coke on the catalyst.

Because the process is dominated by acidic catalysis, if the cracking processes are taken to the extreme, the cracking reactions (known as β-scission) result in propylene and branched olefins such as isobutene. These olefins dominate the light gas products. Ethylene is a very minor component and its presence may be due to a small amount of thermal cracking taking place.

In order to improve the olefin yield, zeolites which are more acidic than zeolite-Y are added to the matrix. These are mainly based on the smaller pored zeolite ZSM-5. This zeolite processes smaller molecules produced by the main cracking process and continues the cracking to smaller olefins and aromatics.

Product Yields

A typical product slate is given in Table 10.1; the data is taken from Maples[4]. The yields are dependent on the nature of the feedstock and the severity of the operation. Because the objective of FCC operation is the production of liquid fuels, particularly gasoline, there is a wide range of practical outcomes for propylene yield. The data in the table is considered typical and shows a propylene yield of 5.5% (wt.) on feed.

Table 10.1: Typical FCC Cracking Yields

	Vol%	Mass %	Mass %
C3	1.56	1.10%	0.99
C3=	8.56	6.07%	5.50
iC4-	4.05	3.26%	2.96
nC4	1.10	0.82%	0.74
C4=	9.94	8.01%	7.26
GASOLINE	54.19	54.99%	49.81
LCO	14.54	18.07%	16.37
HCO	6.16	7.68%	6.95
Gas %			2.99
Coke%			6.43
TOTAL	100%	100%	100.00

Table 10.2: FCC Cracking to Propylene

	T($^{\circ}$C)	Cat/Oil	C3= %	Coke%
Base	532.2	8.5	4.6	5.44
Case 1	543.3	9.1	6.9	5.58
Case 2	548.9	9.4	9.3	5.65
Case 3	554.4	9.6	12.4	5.70

FCC units are very large operations typically taking 30,000 bbl/d of feedstock. This generates about 80,000 t/y of propylene. This is sufficient to give enough propylene for a modest sized polypropylene unit.

A more systematic study has been produced by Golden *et alia*[5] with representative data given in Table 10.2. The table illustrates that higher propylene yield is a consequence of increasing severity in the FCC operation; that is increasing temperature and the catalyst to oil ratio increases the propylene yield. There is a concomitant increase in the amount of coke deposited on the catalyst.

The consequence of this is that since there is a correlation between the propylene yields and coke; higher propylene yields are associated with higher carbon dioxide emissions as the coke is burnt-off in the FCC regenerator.

Fujiyama *et alia*[6] have proposed reconfiguring FCC operations to increase propylene yield. The group have demonstrated a down-flow reactor system operating at high catalyst to oil ratios (>15), high reaction temperature (> 550°C) and short residence time (< 0.5 sec) and obtained propylene yields over 15%.

In the past few years workers at Sinopec have been prominent in developing FCC operations which target propylene as a major product. The increased propylene yield is a function of catalyst developments and increasing the cracking temperature. This variation is known as Deep Catalytic Cracking (DCC) and there are two main variants. Table 10.3 illustrates typical yields that can be achieved[7].

The results indicate that at cracking temperatures of about 550°C, about 20% of the product can be propylene. As well propylene there is a larger portion of ethylene produced. This may be due to either increased

Table 10.3: Propylene Yield Using Special Catalysts and High Temperatures

Refinery	Daqing	Anqing	TPI	Jinan	Jinan
Mode	DCC -1	DCC -1	DCC -1	DCC -1	DCC - II
Feed	Para - VGO[a]	Nap – VGO[b]	Arab VGO DAO + WAX[c]	VGO + DAO[d]	
Temperature (C)	545	550	565	564	530
Ethylene (wt%)	3.7	3.5	5.3	5.3	1.8
Propylene (wt%)	23.0	18.6	18.5	19.2	14.4
Butenes (wt%)	17.3	13.8	13.3	13.2	11.4
of which Isobutene	6.9	5.7	5.9	5.2	4.8
Isopentene					5.9

(a) paraffinic vacuum gas oil, (b) naphthenic vacuum gas oil (c) Arabian vacuum gas oil plus de-asphalted oil plus wax, (d) vacuum gas oil plus de-asphalted oil

Table 10.4: Ethylene and Propylene from Low Value Naphtha Streams

Feedstock Yield (wt%)	FCC Naphtha	Coker Naphtha	Pyrolysis C4	Pyrolysis C5
Fuel gas	13.6	11.6	7.2	12.0
Ethylene	20.0	19.8	22.5	22.1
Propylene	40.1	38.7	48.2	43.8
Propane	6.6	7.0	5.3	6.5
C6+ gasoline	19.7	22.9	16.8	15.6

homogeneous cracking in this system or the catalyst promoting non-classical acid cracking to form ethylene.

Recently a new FCC catalytic system has been proposed which will generate ethylene and propylene from low value olefin rich naphtha feedstock[8]. Typical yields are given in Table 10.4.

In essence this process builds on the ability of zeolites catalysts such as ZSM-5 to establish equilibrium between olefin homologues. Thus when fed a long chain olefin (say octene) at high temperature (typically 500°C), lighter olefins (ethylene and propylene) will be produced. This is essentially the reverse of the olefin polymerisation process which works at lower temperatures to produce polymer gasoline and light diesel from light olefins, such as propylene or butene, using acid catalysts, such as phosphoric acid supported on silica.

Economics of FCC Cracking

We are concerned with the economics of FCC cracking from the perspective of the production of propylene. This is complicated by the fact that the duty of the FCC unit is to maximise the production of gasoline for blending; propylene is only a minor product. Clearly the minimum production cost is the cost of production of gasoline. In this, it should be recalled that many refinery operations use the propylene as feedstock for the production of gasoline either by oligomerisation (poly-gasoline) or by reaction with isobutane (alkylate). From the refiners perspective if a price obtained for propylene is higher than gasoline, this bodes well because the overall production cost of gasoline (the objective) will be reduced. For oil at $70/bbl, the gasoline price approximates to $654/t and this can be taken as the indicative production cost (value) for propylene by FCC.

The cost/benefit of producing additional propylene is then the relative loss of volume in the production of gasoline versus the higher price obtained. This is further complicated in the DCC type operations because there is insufficient data available to attest to the quality of the gasoline and cycle oil by-products.

Fixed-Bed Cracking

One variant of this route is the use of ZSM-5 family zeolites to interconvert olefins; this is broadly similar to the Superflex process. The usual approach is to feed a high olefin (olefinic naphtha) to a fixed-bed catalyst operating at a relatively high temperature (> 400°C). This establishes an equilibrium favouring lighter olefins and in particular propylene. One proposal is to use C4 and C5 olefins to generate ethylene and propylene[9].

Catalytic Cracking to Produce Ethylene

There are several objectives to produce ethylene by catalytic cracking, namely:

- To reduce energy demand

- To reduce greenhouse emissions
- To allow the use of heavier feedstock such as crude oil.

Attempts realise these objectives are based on catalysts able to handle heavy feedstock at relatively low temperatures (550°C versus the 850°C for steam cracking).

However there are several major hurdles. The most common catalysts are based on acid catalysis with Bronsted or Lewis acid sites; these sites promote the formation of propylene rather than ethylene as is witnessed by conventional FCC operations. Ethylene is promoted by free radical processes. Catalysis of free radical reactions is rare, but not unknown[10]. One route is to take a conventional acid catalysis and to neutralise the acid sites with alkaline metals (magnesium, calcium) or phosphorus or a mixture of such. This can generate a further problem, in that the catalyst promotes the formation of carbon (coke) and hydrogen which are thermodynamically favoured at the reaction temperatures.

The higher ethylene yields observed in the DCC type processes has led developments towards the catalytic cracking of heavy oils to ethylene. A typical yield from cracking a gas oil (b.p. 229-340°C) with 45% paraffins, 35.7% naphthenes and aromatics 18.2% is illustrated in Table 10.5[11].

Table 10.5: Catalytic Cracking to Ethylene at 660°C

Ethylene	wt%	21.86%
Propylene	wt%	15.04%
BTD/C4 olefins	wt%	5.70%
Gasoline +	wt%	26.92%
Hydrogen/ethane	wt%	15.52%
Methane	wt%	2.96%
Propane	wt%	0.23%
Butane	wt%	0.23%
Coke and losses	wt%	11.55%
TOTAL	wt%	100.00%

The higher ethylene yield is delivered by a high temperature (660°C). This is high compared to normal FCC type operations, but considerably lower than the temperatures typical for steam cracking

(880°C). In theory this should lead to a lower unit energy demand which may reduce the total amount of carbon dioxide emitted.

Unlike FCC, the process requires a high level of steam included in the cracking unit. This is lower than the amount of steam in steam-cracking, but the presence of the steam at the high reaction temperatures could lead to poor catalyst stability.

The catalyst is based on high levels of a ZSM-5 type zeolite which has been doped with a combination of phosphorus, magnesium and calcium. This type of formulation has been used to produce ethylene and propylene from methanol and is known to promote olefin formation from a wide variety of feeds[12].

There is a relatively large amount of coke formed (11.5%). This means that in practice this technology will require a large regeneration unit, much like that in an FCC operation. It is not clear if this level of coke, which will go on to produce carbon dioxide, will result in lower greenhouse emissions than the conventional routes using higher temperature steam cracking operations.

Catalytic Dehydration of Paraffin to Light Olefins

We are primarily concerned with the production of propylene, butene, isobutene and butadiene from a paraffin of the same carbon chain length. Early technology concentrated on the production of butadiene from n-butane by dehydrogenation over a chromia catalyst – Houdry Catadiene Process. During the 1970s there arose a large market for MTBE (methyl-*tertiary*-butyl-ether) as a gasoline additive. This requires isobutene as a feedstock and the large volumes justified the conversion of n-butane from gas field developments, firstly to isobutane and then the dehydrogenation to isobutene by the Catofin Process, which is variation of the Catadiene Process.

This early process is very capital and maintenance intensive and spurred improvements to catalysts and technology. The Oleflex process (UOP) has been commercialised to dehydrogenate propane to propylene using a platinum supported catalyst. Philips has developed a process using steam as a diluent and uses a tin-platinum catalyst.

Thermodynamics

Under normal pressure, the conversion of paraffins to olefins and hydrogen is not favoured thermodynamically until temperatures are in the region of 900 K (630°C) or higher: Figures 10.2A and B.

The first figure illustrates that ethane and propane dehydrogenation becomes favourable at temperatures over about 900 K under ambient pressures. For the conversion of butane (C_4) to butadiene (BD), higher temperatures are required.

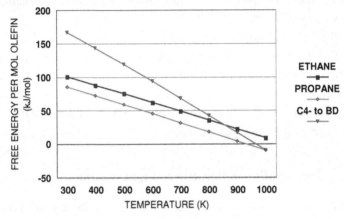

Figure 10.2A: Free energy changes of paraffin dehydrogenation

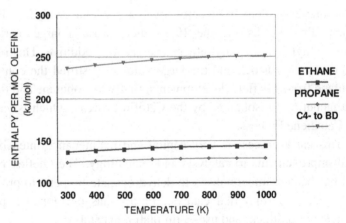

Figure 10.2B: Enthalpy changes of paraffin dehydrogenation

Since the conversion of paraffins to olefins is accompanied by an increase in volume, higher conversions are improved at lower pressures. This is achieved by either using low pressure converters or diluting the feed in a large volume of steam. Furthermore, the reaction is very endothermic as illustrated in the second figure, so a large amount of reaction heat has to be provided.

Houdry Process

The outline of the process is shown in Figure 10.3[13]. The process is described for the conversion of n-butane to butadiene. Typically a C4 feed is heated to the required temperature (typically over 500°C). This is led at low pressure to a series of converters (1) operating in parallel charged with chromia catalyst. These reactors contain hot catalyst from the regeneration step. As the reaction proceeds the catalyst cools and cokes. The catalyst is taken off-line (typically on-line times are 15 minutes). There is then a short period when inert gas (nitrogen) is passed through the catalyst to remove any hydrocarbon present before a blast of air regenerates the catalyst by burning off the coke. This combustion reheats the catalyst to the operating temperature. Another short period of inert gas removes any air that is present before going back on-line.

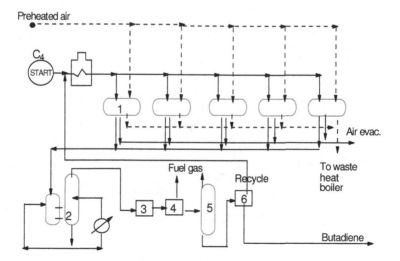

Figure 10.3: Catofin/Catadiene process

Product gases are passed to separators. For butadiene there is a wash step to remove heavy by-products (2), this is omitted in the Catofin versions of the process. The product gases are passed to a compressor (3). The suction-side of the compressor ensures the upstream units operate under vacuum and the compressor exit side raises the pressure to facilitate separation. The compressed gas is separated (4) into the required C4 stream and hydrogen rich gases which are purged from the system. A column (5) removes lighter olefins and (6) separates unconverted feedstock from the butadiene product. The unconverted butane is recycled and the hydrogen and light gases produced from cracking are passed to form a fuel gas. The butadiene process produces a lot of by-product hydrogen, which for optimum economic outcomes should be extracted and used elsewhere or sold.

UOP Oleflex Process

The outline of the process is shown in Figure 10.4[14]. Fresh feed and recycle feed are combined at low pressure and heated to the required temperature (3) and are then passed to a reactor (1). Conversion cools the gases which are then reheated before passing to a second and then a third reactor.

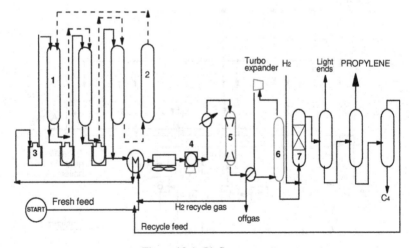

Figure 10.4: Oleflex process

The catalyst progressively cokes and this is handled by continually moving the catalyst through the system until it is finally passed to regenerator unit (2). Regenerated catalyst is passed to the beginning of the process.

The product gases are cooled and compressed (4) to facilitate separation of products and by-products. The suction-side of the compressor ensures that upstream units operate at a low pressure. The product gases are first dried (5) and the cooled product passed to a cryogenic separator (6) which removes hydrogen from the system. Some is recycled with the other portion passed-on for other uses. A selective hydrogenation unit (7) removes dienes and acetylenes. A final distillation train removes light hydrocarbon (C_2.), propylene product, propane, which is recycled, and a C_4 by-product.

STAR Process

In the STAR process (steam active reforming) feed is heated and mixed with steam before passing to the reactor. This avoids the use of pumping to lower the partial pressure of the reactants. The outline of the process is shown in Figure 10.5[15].

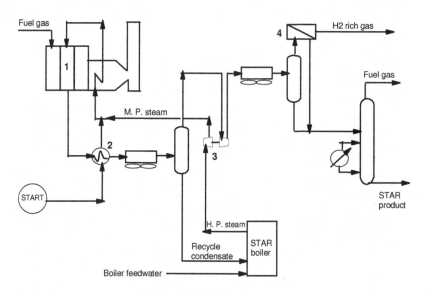

Figure 10.5: Star process

The catalyst (supported platinum promoted with tin) is held in tubes which are held within a furnace arrangement. This avoids catalyst cooling and keeps the catalyst online until coking forces a regeneration step. This is achieved by passing steam through the catalyst to force out hydrocarbon before air is used to burn off the coke. A further steam blast removes air before the catalyst comes back on line. The furnace arrangement is such that some tubes are in operation, some in steaming and some in regeneration mode.

Economics of Paraffin Dehydrogenation

For this analysis the Oleflex technology is used as a base[16]. This has been translated to give the statistics given in Table 10.6. Capital for the production of 350 kt/y of propylene is estimated at $280 million. This requires 412 kt/y of propane. Hydrogen is considered as the only significant by-product which is assumed sold at energy equivalent prices.

Table 10.6: Economics of Propylene Production by Propane Dehydrogenation

	Mt/y	PJ/y	MM$/y	MM$/y
CAPEX			280.24	
OPEX (10% CAPEX)			28.02	28.02
RETURN ON WC (10%)			3.82	3.82
RECOVERY (10%DCF, 20y, FACTOR 0.143)			40.07	
FEEDSTOCK & FUEL PURCHASES				
Feedstock	0.412	20.712	231.37	231.37
Operating Feed (5.5%)	0.023	1.139	12.73	12.73
Fuel (for power)	0.020	0.858	7.15	7.15
Operating Fuel (3%)	0.001	0.027	0.21	0.21
TOTAL	0.455	22.736	251.46	
OUTPUTS				
Propylene	0.350	16.835		424.90
BTD/C4 olefins				
Gasoline				
Hydrogen	0.015	0.769	33.63	33.63
Methane	0		0.00	0.00

Table 10.6 (continued)

Propane				
Butane				
Fuel Oil				
TOTAL	0.365	17.604	33.63	458.53
THERMAL EFFICIENCY (%)		77.4%		
ANNUAL COSTS			323.38	283.31
BYPRODUCT CREDITS			33.63	
UNIT PROPYLENE PRODUCTION COST ($/t)			827.86	
MARGIN				175.2
			$/t	500.6
			c/lb	22.7

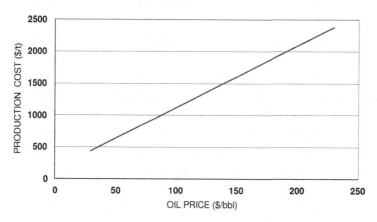

Figure 10.6: Sensitivity of propylene production cost by dehydrogenation to oil price

The production cost is estimated at $827/tonne with an operating margin at 2007 prices at $500/t. The sensitivity to the oil price (which impacts on the price of propane) is illustrated in Figure 10.6.

This figure shows that propylene production cost increases rapidly with increasing oil price. This is based on the assumption that propane is priced according to the world parity price for LPG. In some parts of the world, propane comes from large scale gas developments and is not necessary priced on a world parity basis and this offers lower production cost compared to similar process operations paying parity prices for propane.

Olefin Metathesis

Olefin metathesis offers a means of shifting olefins to olefins with a different number of carbon atoms. Olefin metathesis is the disproportionation or dismutation of olefins over a catalyst, usually based on molybdenum or tungsten. For example, propylene gives ethylene and 2-butene:

$$2C_3H_6 = C_2H_4 + C_4H_8$$

In this case two molecules of propylene form one molecule each of ethylene and 2-butene. Thus, if a petrochemical complex has an excess of propylene (say) this can be converted into ethylene and butene. Similarly, butenes can be used to produce ethylene, propylene and hexene.

The reaction is reversible so that ethylene and butene can be converted into propylene. At present the most common use is to produce additional propylene by reacting butene with an excess of ethylene.

The catalyst has some isomerisation activity so the product linear olefins can have the double bond in any position, similarly any linear isomer can be used as a feedstock. Branched olefins (e.g. isobutene) are not usually converted.

The process layout for production of propylene from ethylene and butene is shown in Figure 10.7[17].

Ethylene and butenes enter the system and are mixed with recycle streams. A guard bed is present to prevent dienes and acetylenes entering the system. The mixed feed is heated and passed to the metathesis reactor which converts ethylene and 2-butene to propylene; 1-butene is isomerised *in situ* to 2-butene. The product is fractionated, first to remove and recycle ethylene and purge lighter gases, and then to produce the polymer grade propylene. Excess butenes are recycled and heavier products removed by a purge.

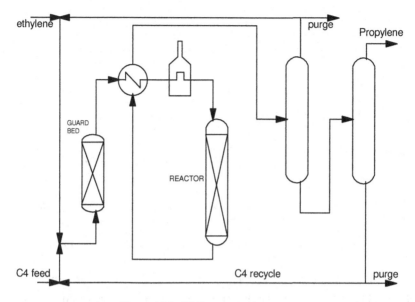

Figure 10.7: Olefin metathesis process

Catalytic Dehydration of Ethanol

Prior to the advent of the petrochemical age, ethylene was made from ethanol by dehydration:

$$C_2H_5OH = C_2H_4 + H_2O$$

Nowadays large volumes of ethanol are made by the reverse reaction, namely acid catalysed hydration of ethylene. However, concern with carbon emissions from other processes and the fact that ethanol is made in very large volumes by fermentation processes, is leading to a new interest in the concept for the production of "renewable" ethylene and hence renewable plastic. The equilibrium of the reaction is shown in Figure 10.8.

The figure illustrates that below 400K (120°C) the equilibrium favours the hydration of ethylene. Higher temperatures favour the dehydration of ethanol with temperatures over 200°C placing the equilibrium well to the favour of ethylene.

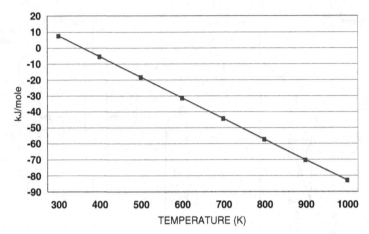

Figure 10.8: Free energy change for ethanol to ethylene and water

The enthalpy of the reaction is +45.3 kJ/mol of ethylene, or 1.62 GJ/tonne. In normal operations this would be delivered by fuel oil or gas, but in totally renewable operations this heat input may be by burning waste produced from the production of starches and sugars used in the fermenting to produce ethanol, e.g. bagasse.

The dehydration reaction is performed over a suitable sold acid catalyst (alumina or silica-alumina) at typically 250°C. The equilibrium is established and the products separated from any unconverted ethanol by distillation. The ethanol is recycled; Figure 10.9.

Ethanol is heated and passed to the converter where the dehydration equilibrium is established. The products are passed to a column which removes the ethylene. Then a second column separates ethanol from water.

Because water is a product and the recycle ethanol will be wet, the ethanol feedstock need not be the highest grade, but instead the easier to produce hydrous ethanol (95% ethanol). If the reaction temperature is low, there should be no contamination from acetylene, which is a problem with higher temperature routes.

Process Economics

The process employs relatively simple unit operations using well known catalysts. Outline statistics are presented in Table 10.7.

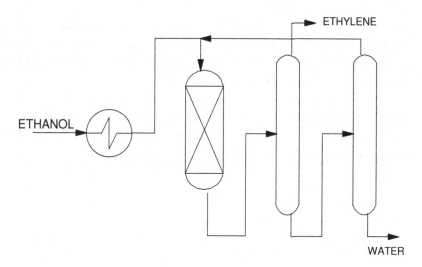

Figure 10.9: Ethanol to ethylene and water

Table 10.7: Statistics for Ethanol Conversion to Ethylene

Ethanol	kt/y	821
Ethylene	kt/y	500
Capital Cost	MM$	150
Fuel Required (85% efficiency)	kt/y	22
Thermal efficiency		93.20%

This order of magnitude estimate is based on a hypothetical plant for the production of 500 kt/year of ethylene with a capital cost of $150 million. This is compared to the cost of a green-field ethane cracker of about $700 million. Process selectivity is assumed 100% with operating allowance for feed and fuel of 5.5% and 3% respectively.

The process requires 841kt or 1041ML ethanol to produce the 500 kt ethylene, this can be compared to the current US fuel ethanol production of about 19,000 ML/y (2007).

The process economics is dominated by the feedstock cost. Because fuel ethanol is widely used as gasoline additive, the cost of ethanol is dependent on the prevailing oil (gasoline) price. At present the price of ethanol is at a premium to gasoline, but as more plants come on stream and absorption into the gasoline pool increases, ethanol may sell at a discount[18]. The fixed variable equation for the production of ethylene from ethanol is shown in Figure 10.10.

With oil at $70/bbl, the price of gasoline is about $655/t. On an energy equivalent basis, ethanol would be valued at $423/tonne. The graph illustrates that with an ethanol price of $500/tonne, the ethylene production cost will be in the vicinity of $970/tonne.

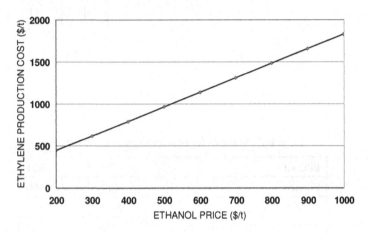

Figure 10.10: Sensitivity of ethylene production cost to ethanol price

Carbon Emissions from Propane Dehydrogenation

The carbon dioxide emissions from propane dehydrogenation is estimated at about $0.9tCO_2/t$ propylene; Table 10.8. The data is derived from the descriptions for the Oleflex process.

This table estimates the carbon dioxide emissions from propane dehydrogenation at about 0.9 t/t of propylene. This should be compared to the value for naphtha which is less than 0.4 t/t. This results from the lower thermodynamic efficiency of the process. However, again emission costs could be lowered by distributing some of the emissions to the hydrogen by-product.

Table 10.8: Carbon Dioxide Emissions for Propane Dehydrogenation

Fuel Used	Mt/y	PJ/y	MtCO2/y
Natural gas			
LPG	0.062	3.107	0.185
Fuel oil	0.021	0.885	0.062
Total	0.082	3.992	0.246
Unit emissions (tCO2/t propylene)			0.704
Flaring			
Propane	0.023	1.139	0.068
Unit emission (tCO2/t propylene)			0.193
Total emissions			0.314
Unit emissions (tCO2/t propylene)			0.897

[1] M. Houdek and J. Anderson, *Hydrocarbon Asia*, Nov./Dec. 2005, p. 34; Jan./Feb. 2006, p. 34

[2] Report reviewed in *Oil & Gas Journal*, Feb. 23, 2004, p. 50

[3] M. Walther, *Oil & Gas Journal*, Jan. 27, 2003, p. 52; S. Golden, R. Pulley and C. F. Dean, *ibid.*, Oct. 4, 2004, p. 44 and P. K. Niccum, M. F. Gilbert, M. J. Tallman and C. R. Santner, *Hydrocarbon Processing*, Nov. 2001, p. 47

[4] R. E. Maples, "Petroleum Refinery Process Economics", PennWell Books, Tulsa, Oklahoma, 1993; Table 12-1

[5] S. Golden, R. Pulley, C. F. Dean, *Oil & Gas Journal*, Oct. 4, 2004, p. 44

[6] Y. Fujiyama, H. H. Redhwi, A. M. Aitani, M. R. Saeed, C. F. Dean, *Oil & Gas Journal*, Sep. 26, 2005, p. 54; *Hydrocarbon Asia*, May/June 2006 p. 20

[7] D. Dharia, W. Letzsch, H. Kim, D. McCue, L. Chaplin; *Hydrocarbon Processing*, April 2004, p.61; see also *Hydrocarbon Asia*, Oct./Sep. 2005, May/June 2006; L. Wang, B. Yang, G. Wang, H. Tang, Z. Li, J. Wei, *Oil & Gas Journal*, Feb. 10, 2003, p. 52

[8] *Hydrocarbon Processing*, "Petrochemical processes 2005", Lyondell SUPERFLEX` Process and ATOFINA/UOP Olefin Cracking Process.

[9] H.V. Bolt and S. Ganz, *Hydrocarbon Processing*, Dec. 2002, p. 77; J. Teng and Z. Xie, *Hydrocarbon Asia*, May/June 2006, p. 26; see also KBR SUPERFLEX Process "Petrochemical Processes 2005", *Hydrocarbon Processing*, ,CD ROM, p.178

[10] I. E. Maxwell, "Non Acid Catalysis by Zeolites" in *Advances in Catalysis*, No. 31, Academic Press, 1982

[11] US Patent, 6, 211, 104

[12] R. J. McIntosh, D. Seddon, *Applied Catal.*, 6, 1983, p. 307-314 and references therein

[13] *Hydrocarbon Processing*, March 1997, p. 116

[14] *Hydrocarbon Processing*, March 1995, p. 104; *Hydrocarbon Processing* "Petrochemical Processes 2005", AAB Lummus Global CATOFIN Process.

[15] *Hydrocarbon Processing*, March 1997, p. 142

[16] The economics of Oleflex technology is described in *Hydrocarbon Processing* of March 2001. The data is scaled to 2007.

[17] *Hydrocarbon Processing*, Petrochemical Processes 2005; ABB Lummus Global OCT process and Axens CCR-Met-4 process.

[18] L. Kumins, *Oil & Gas Journal*, Nov. 26, 2007, p. 18; S.D. Jenson, D.C. Tamm, *Oil & Gas Journal*, May 8, 2006, p. 54

CHAPTER 11

ROUTES TO OLEFINS FROM COAL

In this chapter we consider some commercial routes and emerging technology for the production of olefins from coal. For the most part olefins are made from natural gas and crude oil derivatives - LPG, naphtha, gas oil or residual fuel oil. The cost of these feed-stocks are tied one way or another to the prevailing price of crude oil and the petrochemical operations have to bid for feedstock against the oil-refiners demand for them to produce transport fuels. For example, a major feedstock is ethane. In developed economies the price of ethane is directly linked to the prevailing price of crude oil or indirectly via the natural gas price, which is linked one way or another to the price of crude oil.

Methane (the major component of natural gas) can also be converted into olefins via methanol or the Fischer-Tropsch process. These routes have much in common with the coal to olefins routes in that gas is converted into synthesis gas. Except for comparative production costs the gas based routes are only briefly discussed here[1].

In addition to price considerations, there is the issue of strategic supply. With most of the oil reserves being held in OPEC member countries, particularly in the Middle East, and the major natural gas reserves held jointly between Russia and OPEC, there is concern about supply to many developed countries. By contrast, coal is available across the developed world with major reserves in the USA and China. Even Europe has substantial reserves of coal. Furthermore, the known coal reserves far exceed those of oil and natural gas combined. Indeed, in many coal rich countries, e.g. Australia, prospecting for new coal resources is hardly encountered – the world's coal reserves may be much higher than published statistics.

As a consequence of this, there is now a major focus on the use of coal as a source for fuels and olefins. Coal has a far higher level of carbon than petroleum fuels and natural gas and this inevitably leads to higher emissions of carbon dioxide in the production of fuels and olefins. As a consequence there are research and development projects concerned with the capture of carbon dioxide from coal based operations and the geo-sequestration of the carbon dioxide.

In addition to all this, at the time of writing, there is a major disconnect between the energy price of coal and oil, with coal being much cheaper than oil on an energy equivalent basis. This makes the production of olefins from coal increasingly an attractive option.

Coal to Olefins – Current Technology

Prior to the petrochemical age, ethylene was obtained from coal-gas which was produced by the pyrolysis of coal. Coal pyrolysis was widespread for the production of town gas and is still conducted on a large scale for the production of coke in steel production.

Coke and Town Gas

Technology for the production of coke has been known for many hundreds of years. There are many forms of the process; the two main ones are for the production of coke for iron manufacture and for the production of gas and chemicals.

Carbonization refers to the heating of bituminous coal in ovens or retorts sealed from air to form coke. The process involves thermal decomposition of the coal with distillation of the products. Various technologies are used which perform the process at (i) low temperature (500-750°C), (ii) medium temperature (750-900°C) and (iii) high temperature (900-1175°C). High temperature operation generally favours the production of coke for iron making.

Coke Ovens and Gas Retorts

The coke ovens are held in batteries of many ovens producing coke on a batch basis. A typical coke oven is about 40 ft long, 14ft high with an average width of 17 inches. The coke oven is tapered to facilitate coke removal. The ends of the oven are closed with removable doors. The oven is filled from charge holes at the top of the oven. Volatile products leave the oven from openings in the top which transfer the volatile materials to collecting mains. The coke oven is heated with coke oven gas burned in the oven walls. Typically 35% of the coke oven gas is used in this process. When carbonisation is complete, a pusher machine pushes the hot incandescent coke out of the oven into a receiving truck.

For the production of town gas, the operation is similar. Carbonization is usually performed at a lower temperature and the ovens are smaller and generally referred to as retorts.

After production, the volatile matter is passed to a downstream processing train which removes the products. This is similar for both coke ovens and gas retorts. The general layout is illustrated in Figure 11.1.

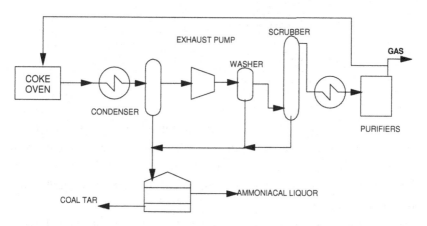

Figure 11.1: Layout for producing coke or town gas

The coke oven or coal retort is heated using a portion of the produced gas. The system for coke ovens generally operates at slightly reduced pressure and the exhaust pump draws the gas and liquids through a condensation system and liquid knock-out. The gases are water washed free of entrained liquids. The liquids comprise a hydrocarbon (tar) phase and an aqueous phase (ammoniacal liquor).

The coal tars are worked up to produce coal chemicals: naphtha, cresols and phenols. The ammoniacal liquor is distilled to produce ammonia (which is often converted into ammonium sulphate).

In carbonisation systems, coke forming in the absence of oxygen restricts the formation of carbon oxides to what can be formed from water and oxygen present in the coal. Some hydrogen is formed by the water gas shift reaction but most (and the methane formed) is a consequence of decomposition of the large coal hydrocarbons into the elements. The typical gas composition is shown in Table 11.1.

Table 11.1: Typical Composition of Coal Gas

	Vol %
Carbon Monoxide	6.8
Hydrogen	47.3
Methane	33.9
Carbon Dioxide	2.2
Nitrogen	6.0
Ethylene etc.	3.8
Fuel Value MJ/m^3	22.0

As can be seen, there is a small portion of ethylene produced in the gas. This small amount of olefin was sufficient for the early days of the chemical industry but soon became displaced by the larger production volume of olefins by steam cracking of ethane, LPG and naphtha from oil and gas sources.

Currently there is active work on new coal pyrolysis technology. These processes are primarily focussed on upgrading low quality coal resources such as lignite. These pyrolysis routes produce a carbon rich solid, which has higher and more useful specific energy density than the feedstock, and a pyrolysis liquid which can be used as a substitute for fuel oil. A portion of the gas produced is used in the pyrolysis and excess is available for other uses. The potential is that these routes could be conducted on a very large scale and significant volumes of pyrolysis gas containing some olefins would be produced.

Indirect Conversion of Coal to Olefins

More olefins are produced by routes which convert the coal into an intermediate product which is subsequently converted into olefins. This could be a liquid, which is then subject to pyrolysis cracking. Most interest focuses on the gasification of coal into carbon monoxide and hydrogen, commonly known as synthesis gas. From synthesis gas olefins can be produced directly or via further intermediates, such as naphtha or methanol.

Coal Gasification

The gasification of coal has been practiced for many years and the subject of major research and demonstration programs. The main aims of this gasification work are the efficient production of synthesis gas for the production of electricity, fuels and chemicals.

The gasification process involves the combustion of coal with a restricted amount of oxygen. For the most part the oxygen is provided by a separate air separation unit (ASU) rather than air which would otherwise introduce a large amount of nitrogen into the synthesis gas. The principal chemical reactions are:

Solid – Gas Reactions:

$$\text{Combustion: } C + O_2 = CO_2$$

$$\text{Steam Carbon: } C + H_2O = CO + H_2$$

$$\text{Hydro-gasification: } C + 2H_2 = CH_4$$

$$\text{The Boudouard Reaction: } C + CO_2 = 2CO$$

Gas Phase Reactions

$$\text{Water-Gas-Shift: } CO + H_2O = CO_2 + H_2$$

$$\text{Methanation: } CO + 3H_2 = CH_4 + H_2O$$

As well as these reactions, large coal molecules undergo pyrolysis and hydro-pyrolysis to smaller molecules.

There are many variations on the gasifier, but most fall into three main categories – moving-bed, fluidised bed and entrained-bed gasifiers. These are shown diagrammatically in Figure 11.2.

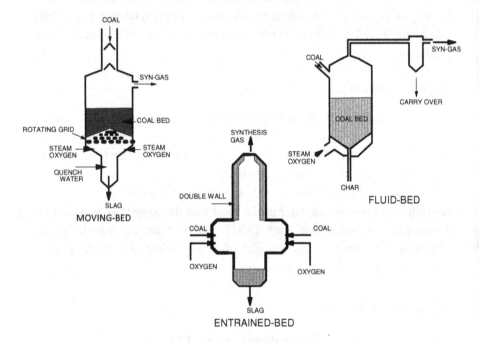

Figure 11.2: Gasifier types

In moving-bed gasifiers, relatively large lumps of coal (6 to 25mm) are added through a hopper system at the top of the gasifier and fall onto a bed of burning coal. Steam and oxygen enter the bottom of the gasifier and move upwards through the burning coals. As it burns the coal moves down the burning bed and comprises only ash when it reaches the bottom.

A feature of this type of gasifier is that the temperature in the bed peaks sharply in a burning (gasification) zone about a third of the way up the coal bed. The maximum temperature is about 1,300°C.

A consequence of this is that fresh coal falling on top of the hot coals undergoes pyrolysis and emits coal gas products – hydrogen, methane, ethylene, light and heavy hydrocarbons and coal tar products – which contaminate the synthesis gas.

Ash falls out of the bottom of the bed through a rotating grate which prevents clogging. Synthesis gas exits a side arm near the top of the gasifier. Older versions of this gasifier were usually non-slagging (i.e. the ash does not melt), however, slagging versions of the gasifier are now available.

In the fluidised-bed gasifier, finely ground coal particles (<5mm) enter the top and are fluidised in upward flowing steam and oxygen, fed from the bottom of the gasifier. Fluidisation results in some carry-over of particles which are separated from the synthesis gas in a cyclone and returned to the gasifier.

Combustion occurs over a larger range in the gasifier which is almost homogeneous in temperature at about 1,000°C. Because of the nature of the fluidised bed there is a relatively high level of unconverted carbon present in the ash. The ash and unconverted coal (char) exit the bottom of the vessel. This gasifier is represented by gasifiers of the Winkler type.

In the entrained-bed gasifier, very fine coal (< 1 mm), sometimes as a water slurry, enters the gasifier and is mixed with steam and oxygen. Combustion is immediate and a very high temperature is achieved (1,800°C). Most of the ash melts and forms a slag over the inner wall of the vessel. The slag runs down the walls into a collector. Some ash is entrained in the synthesis gas which typically exits the top of the reactor. Heat is recovered by a waste heat boiler and the ash is then removed by means of a cyclone.

There are several variants of this type of gasifier. Shell and Uhde separately offer a gasifier based on the Shell-Koppers development with coal being delivered by nitrogen and steam. The Siemens gasifier is similar. The GE gasifier (formerly Texaco gasifier) introduces the coal in water slurry. The Conoco-Philips gasifier is similar. The KBR gasifier uses a complex pipe system to circulate combustion products similar to that in a Fluid Cat-Cracker.

Table 11.2: Performance of Different Gasifiers with Illinois No.6

	Lurgi	BG/L	KRW	Texaco	Shell
Type of bed	Moving	Moving	Fluid	Entrained	Entrained
Pressure (MPa)	0.101	2.82	2.82	4.22	2.46
Ash type	ash	slag	agglom[a].	slag	slag
H_2	52.2	26.4	27.7	30.3	26.7
CO	29.5	45.8	54.6	39.6	63.1
CO_2	5.6	2.9	4.7	10.8	1.5
CH_4	4.4	3.8	5.8	0.1	0.03
Other hydrocarbons	0.3	0.2	<0.01	Nil	Nil
H_2S	0.9	1.0	1.3	1.0	1.3
H_2S/COS	20/1	11/1	9/1	42/1	9/1
$N_2 + A$	1.5	3.3	1.7	1.6	5.2
H_2O	5.1	16.3	4.4	16.5	2.0
$NH_3 + HCN$	0.5	0.2	.08	0.1	0.02

[a] agglomerates

The different gasifier types have differing operating temperatures, pressures and residence times[2]. These factors influence the product slate out of the gasifier. Table 11.2 illustrates the impact of using the same black coal source (after Perry T27-11)[3].

The Lurgi gasifier is the older version of the moving-bed gasifier operating at near atmospheric pressure. This gasifier is "non-slagging". The newer versions of this gasifier are the British Gas/Lurgi and the Sasol/Lurgi gasifiers. These are high pressure gasifiers operating at about 30bar. The coal bed is hotter and the ash forms a molten slag. Both these gasifiers produce significant amounts of methane, other hydrocarbons, ammonia and hydrogen cyanide.

These other products represent carbon lost to products other than synthesis gas. They have to be extracted downstream and disposed of. In the "other hydrocarbons" category are coal tar products – phenol and cresols – and can be extracted and sold as by-product. Otherwise the by-products need to be separated and burned to produce electricity.

The fluid-bed and entrained-bed gasifiers generally operate at higher and more homogeneous temperatures. This eliminates the production of the higher hydrocarbons and reduces the formation of ammonia and hydrogen cyanide. However, the fluid-type still produces a significant amount of methane.

Of the main product gases (H_2, CO, CO_2 and H_2O) the key differences are in the amount of deep oxidation products that are formed (CO_2 and H_2O). The BG/L and the Texaco produce high levels (about 20%) of these gases which suggest relatively high oxygen consumption.

The relative amounts of H_2 and CO (synthesis gas stoichiometric ratio) are immaterial at this juncture since application of water-gas-shift (WGS) moves carbon monoxide to hydrogen or *vice versa* if required.

$$CO + H_2O = CO_2 + H_2$$

Synthesis-gas Clean-Up

The design of the system clean-up operations and the selection of appropriate carbon dioxide extraction technology is dependent on the gasifier type and the amount of non-synthesis gas present which has to be removed.

In a typical process, after production, the synthesis gas is "shifted" to the stoichiometric ratio required for the downstream operation using the WGS process. This is followed by carbon dioxide removal using an acid-gas type solvent extraction system. The block-flow of the system is typically as shown in Figure 11.3.

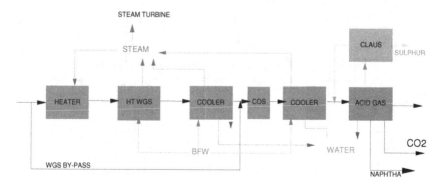

Figure 11.3: Layout for post gasifier synthesis gas clean-up

Sulphur in the coal contaminates the synthesis gas as hydrogen sulphide, this is also removed in the acid gas plant. Carbonyl sulphide (COS) can also be present and this is reduced to acceptable levels by hydrolysis with steam in a unit prior to the acid-gas plant by the reaction:

$$COS + H_2O = CO_2 + H_2S$$

Hydrogen sulphide can be either passed to a Claus unit[4] with the duty to convert the hydrogen sulphide to sulphur or mixed with the carbon dioxide for geo-sequestration. This latter system is in operation at the Dakota Coal Gasification plant in the USA with carbon dioxide and hydrogen sulphide passed by pipeline to a geo-sequestration facility in Canada.

Fischer-Tropsch Process

The Fischer-Tropsch process converts synthesis gas into hydrocarbon products. It was extensively used by Germany in the Second World War and developed in South Africa during the Apartheid years. It is now subject to extensive research and development for the conversion of coal into liquid fuels as an alternative to crude oil. The general process flow-sheet is shown in Figure 11.4.

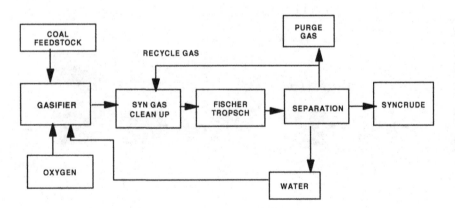

Figure 11.4: General layout for Fischer-Tropsch process

In the Fischer-Tropsch process the coal gasifier produces synthesis gas. This goes to water-gas shift and acid gas removal of carbon dioxide and hydrogen sulphide. The Fischer-Tropsch synthesis converts the synthesis gas into hydrocarbons (-[CH_2]-) and water:

$$CO + 2H_2 = -[CH_2]- + H_2O$$

Some process variants produce carbon dioxide instead of water:

$$2CO + H_2 = -[CH_2]- + CO_2$$

The separator produces a water phase, a hydrocarbon liquid phase (which can be regarded as a synthetic crude oil) and a recycle gas. Part of the synthesis gas is purged to stop the build up of inert materials such as nitrogen. The recycle gas contains light hydrocarbon gases, unconverted synthesis gas and carbon dioxide produced in the process. This is sent to a gas treatment plant for recovery of synthesis gas. This operation may be integrated into the gas clean-up operation of the fresh synthesis gas from the gasifier.

The synthetic crude is passed to a refining section where the various hydrocarbons are separated and processed to produce transport fuels.

The Fischer-Tropsch process has several variants one of which (a high temperature process) produces significant volumes of olefins. This particular variant, known as the Synthol Process, is used to produce fuels from both coal and natural gas in South Africa. A recent development of this process (The Sasol High Temperature Process[5]) has replaced the original entrained-bed reactors with fluid-bed reactors. The product breakdown is shown in Table 11.3.

As well as liquids (gasoline, C_5 -160°C; diesel, 160-350°C) the process produces a large quantity of light olefins, 24.7% of the products. It should also be noted that the gasoline fraction also contains a large quantity of olefins – pentenes, hexenes etc. carbon lost to the water phase include alcohols, ketones, acids and esters.

This high temperature process has been optimised for the production of liquid products. It is feasible that the process could be

Table 11.3: Typical Product Distribution from High Temperature Fischer-Tropsch

	Wt%
Ethylene	4.0
Propylene	11.4
Butenes	9.3
Fuel Gas	17.8
C5 to 160°C	32.5
160-350°C	13.0
> 350°C	5.4
Losses to Water	6.5
TOTAL	99.9
Total Olefins	24.7

further refined and optimised for the olefins, for example by higher temperature operation and lowering the stoichiometric ratio of the synthesis gas.

The high temperature process is the only commercially proven process for the production of olefins and liquids from coal. Current developments favour a low temperature process which is commercially proven to produce liquids and wax from coal or gas. The low temperature process produces a waxy synthetic crude oil which is cracked to produce diesel of high cetane and naphtha. The naphtha, which has high level of linear paraffins, is sold on the petrochemical naphtha market rather than conversion into gasoline. The conversion of this naphtha into olefins by steam cracking has been addressed in previous chapters.

Alpha Olefins

One intriguing aspect of the Fischer-Tropsch process is the production of linear alpha-olefins. These can be separated as inter-mediates in the process and in theory the process could be optimised to produce these valuable products from coal or gas.

Methanol

Another indirect route to olefins is via methanol[6]. For methanol synthesis, a synthesis gas containing some carbon dioxide is acceptable so that a certain quantum can be left in the gas. Figure 11.5 illustrates the route from synthesis gas which is typically tailored to a stoichiometric ratio of 2/1 (H_2/CO) with about 3-4% carbon dioxide left in the feed gas.

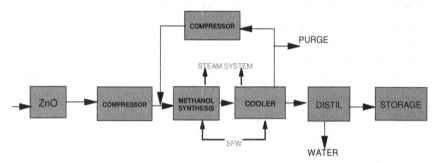

Figure 11.5: Layout for methanol synthesis

Sulphur is detrimental to the synthesis and trace amounts of sulphur are removed using zinc oxide prior to synthesis. After the production of synthesis gas, the methanol synthesis requires compression to about 100bar. The methanol synthesis loop comprises a reactor, a separator and recompression of the recycle gas. A purge gas can be used to produce power supplemented by steam raised in the methanol reactor and the coal gasifier. The crude methanol produced can be upgraded to chemical grade product by distillation. The intermediate methanol is passed into storage. The reaction stoichiometry is:

$$CO + 2H_2 = CH_3OH$$

And

$$CO_2 + 3H_2 = CH_3OH + H_2O$$

To a good approximation, the quantity of water in the raw methanol is given by the second equation relating the water content to the concentration of carbon dioxide in the synthesis gas.

The fact that the methanol is stored as an intermediate brings strength to this route as it de-couples the methanol synthesis from the subsequent conversion of methanol into olefins.

Methanol to Olefins

The conversion of methanol into olefins is similar to the commercially proven methanol to gasoline (MTG) which was commercialised using natural gas as the feedstock in New Zealand. The variant generally uses similar catalysts to produce light olefins only, rather than the iso-paraffins and aromatics of the MTG process. This leads to the prospect of coal or gas conversion into resins (solids). These high value products may be easier to transport and sell than liquid fuels; Figure 11.6 illustrates the basic unit operations for the process.

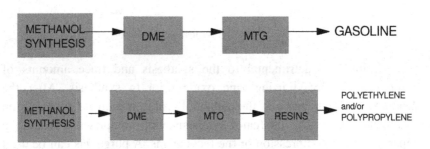

Figure 11.6: Process steps for converting methanol into gasoline and olefins and resins

The methanol, which need not be the highest grade chemical methanol, is produced and stored prior to feeding to the methanol to olefins (MTO) plant. The conversion of methanol into olefins is highly exothermic and in order to help control heat evolution some processes use a primary reactor to convert some of the methanol into dimethyl ether (DME) by the reaction:

$$2CH_3OH = CH_3.O.CH_3 + H_2O$$

The methanol to olefins (MTO) route can be optimized to produce either ethylene and propylene or solely propylene for which there is strong and increasing demand. The basic stoichiometry for ethylene is:

$$2CH_3OH = C_2H_4 + 2H_2O$$

Higher olefins are produced by reaction of ethylene with methanol:

$$CH_3OH + C_2H_4 = C_3H_6 + H_2O$$

However, in detail the conversion of methanol into olefins is quite complex.

Early Mobil Methanol to Olefins Processes (MTO)

Early attempts to convert methanol into olefins were based on the zeolite ZSM-5. The Mobil MTO process was based on the fluidised bed version of the MTG technology. Conversion took place at about $500°C$ allegedly producing almost complete methanol conversion. However, careful reading of the patent literature indicates that complete methanol conversion may not have been achieved by this means. Because of incomplete conversion, there would be a necessity to strip methanol and dimethyl ether from water and hydrocarbon products in order to recycle unconverted methanol. In this variant, the total olefin yield is less than 20% of the products of which ethylene is a minor but not insignificant product. The major product is gasoline. Ethylene is difficult to process and has to be treated specially. Claims that it is possible that ethylene can be recycled to extinction conflict with the known behaviour of ethylene in zeolite catalyst systems and have to be viewed with some suspicion.

The MTC process was primarily designed to produce ethylene by operating a MTG type catalyst and process at low pass conversion in a fixed bed reactor. The route was developed by A.E.C.I. in South Africa who demonstrated the process to pilot plant scale.

The principal reaction is brought about at low conversion in a series of reactors, (10% conversion per reactor with ca. 40% conversion overall). The products, both aqueous and hydrocarbon phases, are heavily laden with methanol and dimethyl ether and as a consequence extensive extraction and recycle is required.

The principal product is ethylene. The higher products are rich in olefins (66% olefins in C_3 + C_4 which are 41% of total). Like Mobil MTO, this process also produces a good quality gasoline and a heavy gasoline which may require hydro-treatment prior to use.

UOP Methanol to Olefins Process

The UOP process, developed jointly with Norsk Hydro/Statoil[7], and has been developed to semi-commercial scale in Norway. The process uses proprietary catalysts based on a SAPO molecular sieve.

Two variants of the process are available, one maximising ethylene and the other propylene. The performance appears to be similar to that of the conversion of methanol to olefins using small pore zeolites. Such systems suffer from high methane yield (which has to be recycled back to a reformer) and high coke yields. The formation of olefins is promoted by using crude methanol, which can contain up to about 17% water.

The coke formation leads to catalyst fouling. This is solved in the UOP Process by continuously removing a portion of the catalyst and passing this to a separate regenerator. After regeneration by combustion of the coke in air, the catalyst is sent back to the main reactor. In concept this is similar to fluid-cat cracking of refinery stocks. The process layout is illustrated in the Figure 11.7.

After separation of the mixed olefins the product work up is similar to that in a steam cracker using LPG feedstock. Small amounts of carbon dioxide are removed and the hydrocarbon gases are dried before passing to a de-ethaniser column. The C_2- fraction is passed to an acetylene removal unit before methane is removed from the C_2 stream. This comprises 98+% ethylene, the remainder being ethane. The C_{3+} stream is split between the C_3 fraction (98% propylene) and C_{4+} . The work up of the C_4 stream to produce linear butenes (not shown in the figure) is likely to be less problematic than the corresponding C_4 stream from steam crackers, which is highly complex and cannot be separated by fractionation alone. The process produces little product above C_5.

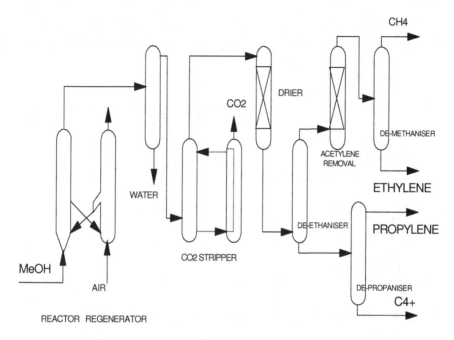

Figure 11.7: UOP MTO process scheme

The Lurgi Methanol to Propylene (MTP) Process

The process has been demonstrated on a pilot scale by Lurgi and Statoil. Sufficient propylene has been produced to make polypropylene resin product by Borealis[8]. This process appears to use an oxide doped ZSM-5 zeolite catalyst in fixed bed reactors. The oxide doping promotes the methanol conversion to olefins. All olefins, other than propylene, are recycled to extinction or purged as fuel gas or produced as naphtha. The flow sheet is illustrated in the Figure 11.8.

Because fixed bed reactors are used, the heat of reaction must be removed. This is achieved by firstly converting some of the methanol to DME in a first reactor (similar to MTG) and then splitting the feed to a series of reactors. Overall, the method resembles the operation of a methanol quench converter where fresh feed is introduced at different points within a single reactor. Operation is at about 500°C at which temperature propylene is favoured over ethylene. Overall promotion of olefin yield is obtained by adding steam. Downstream of the reactors are separation columns, which separate the C_3 product (ca. 80% propylene) from naphtha and fuel gases.

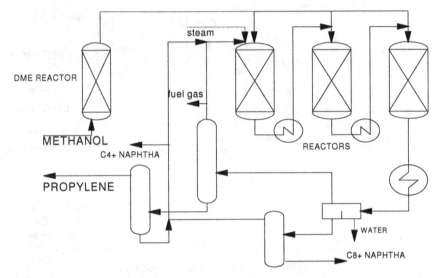

Figure 11.8: Lurgi MTP process

Comparison of Alternative Routes

A comparison of the yield of olefins from the routes discussed above is shown in Table 11.4.

The methanol to gasoline process (MTG) is aimed at producing gasoline (C_5 to 160°C cut). In the process, the light olefins are generally recycled or extracted as LPG. The MTC process (methanol to chemicals) produces high yields of light olefins and aromatic naphtha which can be worked-up to extract aromatics. This product slate is very similar to that for naphtha cracking. The product slate for the UOP/Statoil MTO process can be swung between ethylene rich and propylene rich products. Liquid products are greatly reduced, but there is carbon loss to coke. In the Lurgi methanol to propylene process (MTP), all of the light olefins are recycled to extinction, but this increases the amount of fuel gas and naphtha product. For comparison the high temperature Fischer-Tropsch (HT-FT) process is included.

Table 11.4: Comparison of Olefin Producing Processes

	MTG	MTC	UOP MTO	UOP MTO	MTP	HT-FT
Ethylene	3.2	25.2	45.6	33.6	0.0	4.0
Propylene	4.7	16.5	29.6	44.6	67.9	11.4
Butenes	8.3	5.0	9.5	12.8	0.0	9.3
TOTAL OLEFINS	16.2	46.7	84.7	91.0	67.9	24.7
Fuel Gas	21.2	15.6	5.6	2.0	6.1	17.8
C_5 - 160°C	58.0	33.0	5.5	5.5	26.0	32.6
160 - 350°C	5.0	1.0				13.0
>350°C						5.4
water phase or coke		3.7	4.2	1.5	0.0	6.5

Economics of Olefin Production from Coal and Gas

The route via methanol is analysed. The economics of olefin production for coal and gas is considered in two parts, first the production cost of methanol and then the conversion of methanol into olefins.

For methanol, three scenarios are considered: two large scale gas plants and one from coal. The estimate is made for the production of AA grade which is not usually necessary for the conversion to olefins. This may save a modest amount (5%) of the capital cost. The statistics are given in Table 11.5.

Table 11.5: Statistics for Methanol Production

Feedstock		GAS	GAS	COAL
Production	kt/y	850	1700	1424.9
CAPEX	MM$	428.29	511.48	1193.24
Construction period	years	3	3	4
Plant life	years	15	15	20
Return on capital	%/y	16.34	16.34	15.15
Non feed operating cost	MM$/y	104.62	126.71	258.27
Gas or coal usage	PJ/y	32.16	64.32	45.54
By-products credits	MM$/y	0	0	11.55

A typical world scale plant produces 850 kt/y (2,500 t/d) methanol and requires about 32 PJ/y of gas[9]. Typical construction period is three years with a lifetime of 15 years. Recently, some plants have been constructed at double this capacity (5,000 t/d) and claim much reduced capital costs, which is detailed in the second column. However, although economy of scale applies, some of the reduction in capital claimed probably comes from importing oxygen into the complex to run an auto-thermal gasifier. The coal case produces about 1.4 million tonnes of methanol and is based on the optimum size of an entrained-bed gasifier and requires 45PJ (about 1.8 million tonnes) of black coal. The coal option produces by-products of sulphur, ash and electricity. The fixed variable relationship plotted in Figure 11.9.

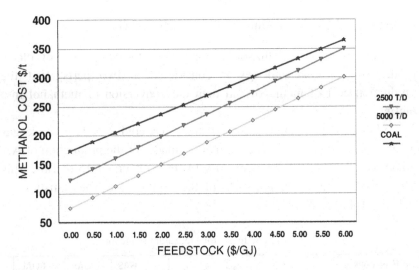

Figure 11.9: Sensitivity of methanol production cost to feedstock price

A typical long-term average traded price for methanol is in the vicinity of $150/tonne. For this, gas prices have to be below $2/GJ and preferably in the vicinity of $1/GJ. At this time, a typical traded price of gas is in the range of $5/GJ and this has stressed many operations which are force to pay this price. For this reason new gas-based world scale methanol plants have migrated to regions of low gas prices in the Middle East.

Table 11.6: Statistics for Conversion of Methanol into Olefins (MTO)

		kt/y
Methanol Used	100.0%	1700
ethylene	19.9%	338.967
propylene	13.0%	220.3285
butenes	4.2%	70.61812
naphtha	2.4%	40.95851
fuel gas	2.5%	41.80592
TOTAL Products	41.9%	712.678

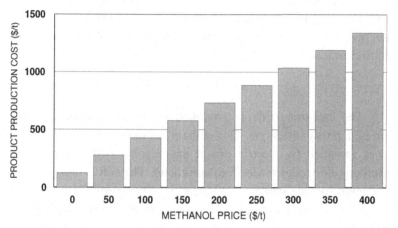

Figure 11.10: Sensitivity of olefins production cost to methanol price

Coal based plants have a higher relative capital cost and the fixed variable curve is at a higher level than for gas based operations. However, coal is widely available at $0.5/GJ (c. $10/t) or less which reduces the production cost to below $200/t.

The fixed variable relationship for the conversion of methanol into olefins using the statistics shown in Table 11.6 is shown in Figure 11.10. The analysis is based on a capital cost of for the methanol to olefin step of $300 million (2007).

This analysis is sensitive to the prevailing price of crude oil and by-products (butenes and naphtha) which change in relation to the oil price. With oil at $70/bbl, which is used as the basis for the sensitivity, the by-products, butenes and naphtha, are valued at $600/t. The fixed variable relationship shows that if methanol is available below \$200/t then the olefin production cost is below about \$750/t.

Coal to Acetylene and Olefins

Coal can be converted into acetylene via calcium carbide. This is used to produce vinyl chloride monomer (VCM) on a large scale in China. In theory, selective reduction of acetylene could give a route to ethylene for coal.

The production of VCM for PVC manufacture proceeds in three steps. The first is the production of calcium carbide by the electrolytic reduction of coke (produced from coal) and calcium oxide in an electrochemical cell.

$$CaO + 3C = CaC_2 + CO$$

The calcium oxide is produced immediately prior to reduction from high purity limestone. This enthalpy of the reaction is +465.6kJ/mol and is provided by electric power and results in the consumption of Soderberg electrodes made from anthracite. The cell is tapped to release the molten carbide which is produced in 80% purity. The off-gas from the cell is typically 80% carbon monoxide and about 10% hydrogen. Following the production of calcium carbide, acetylene is produced by addition of water to the carbide:

$$CaC_2 + H_2O = C_2H_2 + Ca(OH)_2$$

This reaction is exothermic (125.1kJ/mol) and produces about 308kg of acetylene per tonne of 80% carbide. Acetylene is then converted into VCM by addition of hydrogen chloride:

$$C_2H_2 + HCl = CH_2:CHCl$$

The economics of this process is dependent on the availability of low cost coal for the production of carbide and power. The production of 80% calcium carbide requires the resources detailed in Table 11.7; using approximate costs, the cost of acetylene production is estimated at \$681/t.

Table 11.7: Carbide and Acetylene Production from Coal

Carbide is 80% CaC2			$/unit	$/tCaC2
Calcium Oxide	Kg/t	950		
Lime stone	Kg/t	1484	20	29.6875
Coke	Kg/t	550	50	27.5
Electrode C	kg/t	30	100	3
Power	MWh/t	3	50	150
				210.1875
CARBIDE to ACETYLENE	kg/kg	0.308	681.34	

Carbon Emissions for Gas and Coal to Olefins

By far, the major portion of carbon dioxide emissions in the MTO route is in the production of methanol from either gas or coal. The gas route is less carbon intensive as is illustrated in Figure 11.11, which plots the increase in methanol production cost against carbon price.

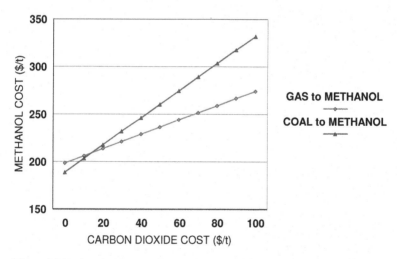

Figure 11.11: Sensitivity of methanol production cost to carbon emission cost

The base statistics for the lines in the figure are for a gas price of $2/GJ and a coal price of $0.5/GJ. These give production costs in the vicinity of $200/t for both cases in the absence of carbon charges.

The graph shows that a $40/tonne carbon dioxide charge will increase the production cost from coal by approximately 25% to around $250/t. For the same impact on gas based operations, a carbon dioxide cost of $70/t will be required.

---

[1] For further information on gas to olefins see D. Seddon, "Gas Usage and Value" PennWell, Tulsa, Oklahoma, 2006

[2] G. Bhandarkar, *Hydrocarbon Asia*, Nov./Dec. 2001, p. 46

[3] Perry's Chemical Engineer's Handbook, 7th Edition (Perry & Green, eds.), McGraw Hill 1997

[4] B. Zarenezhad, *Hydrocarbon Processing*, Oct. 2008, p. 109

[5] T. Chang, *Oil & Gas Journal*, Jan. 10, 2000, p. 42

[6] M. Sutton and P. Roberts, *Hydrocarbon Processing*, July 2007, p. 89

[7] "Petrochemical Processes 99", *Hydrocarbon Processing*, Mar. 1999, p. 125

[8] Borealis A/S, *Press Release* Sep. 16, 2003, "First polypropylene product made from natural gas"

[9] D. Seddon, "Gas Usage and Value" PennWell, Tulsa, Oklahoma, 2006

CHAPTER 12

CLOSING REMARKS

Production Economics

From the above analysis of the various competing technologies and feedstock we can make several observations:

- It is clear that economies of scale count so that petrochemical cracking operations have progressively increased over the years. There seems to be no technical reason why this should not continue over the coming years. A variation on this is the progressive increase in the scale of operation of existing sites, with expansions and de-bottlenecking operations of existing plants going on continuously in many parts of the world.
- The preferred feedstock is ethane, obtained as a by-product to large natural gas operations either for mass delivery into pipeline networks or to service the ever expanding LNG industry. In many jurisdictions ethane from gas is related to the value of the gas and is not directly connected to the prevailing price of crude oil. To-date, this generally leads to ethane based plants producing ethylene, and hence polyethylene resins, at significantly lower cost than liquid cracking operations.
- The favoured position of gas-priced ethane as feedstock has led to a marked increase in the cracking operations based in the Middle East. There appears no reason for this trend to cease.
- Natural gas liquids, such as propane and butane, have a value which is now clearly linked to the prevailing price of crude oil. This has meant that propane and butane cracking are restricted to special situations in time and place when the feedstocks are in excess and prices are depressed relative to their long term relationship to oil.

- The problem for all gas based (ethane, propane, butane) cracking operations remains the very low or poor production of propylene. It seems inevitable that if more ethylene is produced from gas feedstock, more propylene will have to be made from other sources and by other routes. This means a continued growth in catalytic processes for the conversion of paraffins into propylene. Of the diverse range of technologies available, the continued growth in refinery produced propylene by FCC operations would appear the best route, though again propane dehydrogenation will have a role in specific situations.
- Naphtha is likely to remain the main feedstock for petrochemical operations because of the ability of the feedstock to not only deliver both ethylene and propylene, but also BTX aromatics from the rapidly growing fires and speciality polymers markets (nylons, polyester etc.).
- A problem with the use of naphtha is that with all of the downstream plants required, the capital cost for new world-scale operations is very large. However, as demonstrated by the establishment and growth of the integrated complex on Singapore's Jurong Island, this cost can be spread across many organisations.
- When available, the use of waxy residual fuel as a feedstock for cracking will continue. This is a consequence of the price differential between it and naphtha. This price differential compensates for the increased downstream processing costs associated with cracking of heavier feedstock.
- Coal based routes are likely to be established over the next decade. This is predicated on the availability of low cost coal in mainly the world's major economies.

Carbon Emissions

From the above analysis of the various competing technologies and feedstock we can make several observations:

- Carbon emissions are inversely proportional to the thermal efficiency of the process. Minimisation of carbon emissions for existing operations revolves around the mundane tasks of good housekeeping and maintaining all process operations at or better than design capacity.

- From the above, the lighter the feedstock the poorer the thermal efficiency and therefore the higher the carbon emissions. This is counter to the move to increase ethylene production from ethane from natural gas operations.

- Dehydrogenation routes to propylene also increase the amount of carbon emissions relative to the production of propylene from naphtha. Increasing propylene output from FCC operations also increases emissions. Although this is the case for a standalone facility, it is not clear if a full cradle-to-grave analysis would ameliorate or exacerbate the emissions relative to naphtha cracking.

- *Prima-facie* the simplest way to decrease carbon emissions is to use by-product hydrogen as a fuel in the cracking furnace. This denies the use of hydrogen to downstream uses. Generally downstream involves relatively high-added value operations and the use of hydrogen in this way, is in general counter to optimum production economics. Nevertheless, for some operations hydrogen production is well in excess of the downstream needs and recycling to furnace gas would make a significant contribution to reducing emissions.

- A less effective, but more economically viable method, would be to recycle all low-value hydrocarbon by-products to the cracker furnace. This particularly focuses on methane which within the confines of an operation is typically valued relative to the fuel oil price. However, this equally applies to ethane and propane which are generally recycled to the feedstock side of the cracking furnace. Depending on the relative value, it may be optimal for minimising carbon emissions in some operations to use ethane as a fuel rather than a feedstock.

Catalyst Issues

From the above analysis of the various competing technologies and feedstock we can make several observations:

- Thermal cracking operations are not catalyst intensive. However, the use of special coatings to prevent the formation of carbon in the furnace tubes and downstream inter-changers is important. The thermodynamics of the process favours coke formation and it will be important to constantly improve the current coatings and additives used to prevent unwanted reactions in the cracker.

- For downstream cracking operations, the main catalytic process of interest is the selective hydrogenation of acetylene and related compounds. The process is considered to be selective and to only form ethylene, but this could be improved because there is some evidence (including the formation of green oil) that the process is not as selective as generally claimed. It is not clear that the small amount of acetylene present is in fact reduced to ethane rather than ethylene. It is clear there is some room for improvement.

- A more effective acetylene hydrogenation catalyst to ethylene would also facilitate the development of coal to ethylene via the acetylene route, which is at present restricted to the use of acetylene for the production of vinyl chloride.

APPENDICES

A1: Abbreviations and Unit Equivalents

A1.1: Abbreviations

Table A1.1: Abbreviations

ASU	Air separation unit
AVTUR	Aviation turbine fuel - Jet fuel
a	annum (year)
bbl	petroleum barrel
bbl/d	barrels per day
bcfd	billions of cubic feed per day
BTU	British Thermal Unit
BTX	benzene, toluene and xylene mixture
C	Degrees Centigrade (Celsius)
Cf	cubic foot
Cif	container, insurance and freight (destination port price)
Cm	cubic meter
DME	Dimethyl ether
E/P	ethylene to propylene production ratio
F	Degrees Fahrenheit
Fob	free on board (embarkation port price)
GJ	Gigajoule
HDPE	high density polyethylene
HHV	higher heating value (gross)
HP	horse power
K	Degrees Kelvin (absolute temperature scale)
K-factor	UOP or Watson paraffinicity factor
kt/y	thousand metric tonnes per year
kW	kilowatt
kWh	kilowatt hour
L	Litre
Lb	Pound
LDPE	low density polyethylene
LHV	lower heating value (net)
LLDPE	linear low density polyethylene

Table A1.1 (continued)

LPG	Liquefied petroleum gas (usually propane and butane)
Mcf	Thousand cubic feet
MM$	million US dollars (2007)
MMBTU	million (US Customary) BTU
Mt	million metric tonnes
NGL	natural gas liquids
PJ	peta joule (10^{15} joules)
PONA	paraffins, olefins, naphthenes and aromatics
PP	polypropylene
PVC	polyvinylchloride
R	Degrees Rankin (absolute temperature scale in $^{\circ}$F)
S.I.	Système International d'Unités; metric units
T	metric tonne
t/y	metric tonnes per year
VCM	Vinyl chloride monomer
VGO	vacuum gas oil

A1: Unit equivalents

Table A1.2: Equivalents Between S.I. and US Customary Units

S.I	Unit	US Customary
kilo (k)	thousand	M
Mega (M)	million	MM
Giga (G)	10^9	Billion
Tera (T)	10^12	Trillion
Peta (P)	10^15	Quadrillion

A2. Some Useful Conversion Factors for Fuels and Products

Table A2.1: Basic Conversion Factors

FROM		TO
Cm	35.315	cf
cm@15C	35.383	cf@60F
GJ	0.9478	MMBTU
$/GJ	1.055	$/MMBTU
1 kWh	3.6	MJ
kg	2.2046	lb
HP	0.7457	kW
tonne (metric)	1.102	ton (short)

Table A2.2: Temperature Conversions

	C	F
Absolute zero	-273.15	-459.67
Normal	15	59
STP (metric)	0	32
Standard	15.56	60

Table A2.3: Specific Volumes and Heating Values of Liquid Fuels

	L/t	bbl/t	HHV GJ/t
ETHANE	2654		
PROPANE	1998		
BUTANES	1928	12.13	49.6
NAPHTHA	1534	9.00	48.1
GASOLINE	1360	8.56	46.4
AVTUR	1261	7.93	46.4
DIESEL	1182	7.43	45.6
FUEL OIL (LS)	1110	6.98	44.1
CRUDE OIL (35API)	1177	7.40	45
CRUDE OIL (40 API)	1212	7.62	

Table A2.4: Energy Values of Some Products and Intermediates

	HHV (GJ/t)	LHV (GJ/t)
CARBON MONOXIDE	10.1	10.1
BUTENES	48.1	45
ETHYLENE	50.3	47.2
HYDROGEN	141.8	120
METHANOL	22.7	19.5
PROPYLENE	48.9	45.8
DME	31	28.4
CARBON	32.8	32.8
AMMONIA	22.5	18.6

Table A2.5: Properties of Some Coals

TYPE		Wyoming	Witbank	Illinois No 6	Wyodak	German Brown
Ultimate Analysis (DAF)						
Carbon	wt.%	74.45%	81.25%	78.10%	75.60%	67.50%
Hydrogen	wt%	5.10%	5.00%	5.50%	6.00%	5.00%
Oxygen	wt%	19.25%	10.00%	10.90%	16.80%	26.50%
Nitrogen	wt%	0.75%	2.50%	1.20%	0.70%	0.50%
Sulphur	wt%	0.45%	1.25%	4.30%	0.90%	0.50%
		100.00%	100.00%	100.00%	100.00%	100.00%
Ash (as received)	wt%			12.0%	5.9%	6.4%
Moistrure (as rec)	wt%			6.5%	35.0%	5.0%
As received Basis						
Carbon	wt.%			65.91%	53.66%	60.59%
Hydrogen	wt%			4.64%	4.26%	4.49%
Oxygen	wt%			9.20%	11.92%	23.79%
Nitrogen	wt%			1.01%	0.50%	0.45%
Sulphur	wt%			3.63%	0.64%	0.45%
Ash (as received)	wt%			10.13%	4.19%	5.75%
Moistrure (as rec)	wt%			5.49%	24.84%	4.49%
				100.00%	100.00%	100.00%
LHV (as received)	GJ/t			25.80	17.16	9.90
HHV (as received)	GJ/t			26.82	18.10	10.89
LHV (DAF)	GJ/t			30.57	24.18	11.03
HHV (DAF)	GJ/t	29.6	31.4	31.79	25.50	12.13

A3. Cost of Utilities

Table A3.1: Utility Costs

Days per year		340
Hours per year		8160
Electricity		
Purchases	c/kWh	5.0
Export	c/kWh	3.0
Steam		
High pressure	$/t	2.04
Medium pressure	$/t	1.81
Low pressure	$/t	1.36

A4. Nelson-Farrar Cost Indices

Table A4.1: Nelson Farrer Refinery Cost Indices

YEAR	MATERIAL	EQUIP.	LABOUR	INDEX	NF FACTOR
weight	0.4	0.0	0.6		
1946	100.0	100.0	100.0	100.0	21.0670
1947	122.4	114.2	113.5	117.1	17.9968
1948	139.5	122.1	128.0	132.6	15.8876
1949	143.6	121.6	137.1	139.7	15.0802
1950	149.5	126.2	144.0	146.2	14.4097
1951	164.0	145.0	152.5	157.1	13.4099
1952	164.3	153.1	163.1	163.6	12.8787
1953	172.4	158.8	174.2	173.5	12.1438
1954	174.6	160.7	183.3	179.8	11.7156
1955	176.1	161.5	189.6	184.2	11.4370
1956	190.4	180.5	198.2	195.1	10.7992
1957	201.9	192.1	208.6	205.9	10.2307
1958	204.1	192.4	220.4	213.9	9.8499
1959	207.8	196.1	231.6	222.1	9.4862
1960	207.6	200.0	241.9	228.2	9.2326
1961	207.7	199.5	249.4	232.7	9.0525
1962	205.9	198.8	258.8	237.6	8.8651
1963	206.3	201.4	268.4	243.6	8.6496
1964	209.6	206.8	280.5	252.1	8.3553
1965	212.0	211.6	294.4	261.4	8.0581

Table A4.1 (continued)

1966	216.2	220.9	310.9	273.0	7.7163
1967	219.7	226.1	331.3	286.7	7.3491
1968	224.1	228.8	357.4	304.1	6.9281
1969	234.9	239.3	391.8	329.0	6.4026
1970	250.5	254.3	441.1	364.9	5.7740
1971	265.2	268.7	499.9	406.0	5.1887
1972	277.8	278.0	545.6	438.5	4.8046
1973	292.3	291.4	585.2	468.0	4.5011
1974	373.3	361.8	623.6	523.5	4.0244
1975	421.0	415.9	678.5	575.5	3.6606
1976	445.2	423.8	729.4	615.7	3.4215
1977	471.3	438.2	774.1	653.0	3.2263
1978	516.7	474.1	824.1	701.1	3.0047
1979	573.1	515.4	879.0	756.6	2.7843
1980	629.2	578.1	951.9	822.8	2.5603
1981	693.2	647.9	1044.2	903.8	2.3309
1982	707.6	662.8	1154.2	975.6	2.1595
1983	712.4	656.8	1234.8	1025.8	2.0536
1984	735.3	665.6	1278.1	1061.0	1.9856
1985	739.6	673.4	1297.6	1074.4	1.9608
1986	730.0	684.4	1330.0	1090.0	1.9328
1987	748.9	703.1	1370.0	1121.6	1.8784
1988	802.8	732.5	1405.6	1164.5	1.8091
1989	829.2	769.9	1440.4	1195.9	1.7616
1990	832.8	797.5	1487.7	1225.7	1.7187
1991	832.3	827.5	1533.3	1252.9	1.6815
1992	824.6	837.6	1579.2	1277.4	1.6493
1993	846.5	842.8	1620.2	1310.7	1.6073
1994	877.2	851.1	1664.7	1349.7	1.5609
1995	918.0	879.5	1708.1	1392.1	1.5133
1996	917.1	903.5	1753.5	1419.0	1.4846
1997	923.9	910.5	1799.5	1449.2	1.4537
1998	917.5	933.2	1851.0	1477.6	1.4258
1999	883.5	920.3	1906.3	1497.2	1.4071
2000	896.1	917.8	1973.7	1542.7	1.3656
2001	877.7	939.3	2047.7	1579.7	1.3336
2002	899.7	951.3	2137.2	1642.2	1.2829
2003	933.8	956.7	2228.1	1710.4	1.2317
2004	993.8	1112.7	2314.2	1833.6	1.1489
2005	1179.8	1062.1	2411.6	1918.8	1.0979
2006	1273.5	1113.3	2497.8	2008.1	1.0491
2007	1364.8	1189.3	2601.4	2106.7	1.0000

A5. Location Factors

Location factors developed through US DoE Studies[1] *relative* to US Gulf.

Table A5.1: Location Factors

	1	2	3	4
Climate/Terrain	Benign	difficult	difficult	extreme
Gas Transmission	Present	present	no	no
Fresh Water	Present	present	no	no
Ship Loading	Present	present	no	no
Employee Housing	Present	present	no	no
Labour Costs	Low	high	high	high
Relative Capex	1.000	1.155	1.562	2.250
Relative Opex	1.000	1.139	1.520	2.039
Examples	US Gulf	Urban Australia	Remote FE	Offshore
	Canada	New Zealand	Remote Aus	Arctic
		Developed FE		Papua New Guinea
		Middle East		

A6. Methodology for Economic Analysis

What is required is a rapid approach to the determination of the economic viability of a particular technology of interest, namely a concept analysis where speed is not gained at the expense of accuracy. This requires a systematic approach in which various technologies and approaches are treated in the same manner so that the economics from one route to olefins can be compared to another.

The methodology described was devised by ICI PLC in order to evaluate all of the diverse routes to the production of ethylene from any feedstock using widely disparate technologies with different plant construction periods and lives of operation. The methodology has been published by Stratton *et alia*[2] and is generally applicable for energy intensive industries. The basic economic equation is:

$$P = F + C + O$$

Where P, the unit production cost of the production of interest (ethylene say), is equal to the sum of the unit feedstock costs (F), the unit capital costs (C) and the unit non feedstock operating costs (O). This can be expressed as a fixed-variable equation with the fixed part of the equation representing the return on capital (the unit capital costs, C, independent of tax considerations) together with all the unit non-feedstock operating costs (O).

Capital Costs $(C)^3$

The capital costs are developed for green-field projects completely isolated from other facilities. All the costs associated with utilities (unless otherwise accounted) are allowed for in the capital cost. Some processes require small amounts of power. This is considered as an import.

Capital is estimated using published information, and using the location factors and Nelson-Farrer Indices given above, it is adjusted to the US Gulf site and 2007 costs for all processes. Scaling uses the exponent method namely:

Capital of Plant [1]/Capital of Plant [2]
    = {Capacity of Plant [1]/Capacity of Plant [2]}^{n}

where n is a constant with a value which is typically 0.7.

Capital Recovery Factors

For a plant with a capital cost of C_o, the plant investment cost, C, capitalises the return on investment during construction of the plant – it takes account of expenditure and the required return during the construction period.

$$C = C_o \sum_{s=0}^{p} a(s)(1+i)^{p-s}$$

where a(s) represents the breakdown of capital expended over the construction period; p is the first year of production; and s is a general year of the project starting with $s = 0$, with construction complete at $s = p$. The return on investment is i. The values of a(s) are given in Table A6.1.

Table A6.1: Values for a(s)

Construction. Period (years)	1	2	3	4	5
a(0)	100%	50%	30%	17%	4%
a(1)		50%	45%	32%	14%
a(2)			25%	26%	32%
a(3)				25%	36%
a(4)					14%

The general DCF equation can be written:

$$C = \sum_{r=1}^{N}((Rr - FCr - VCr)/(1+i)^r)$$

where r is the production year, with N the final production year and Rr is the total product revenue in year r, FCr is the fixed costs in year r, VCr is the variable cost in year r.

This equation is simplified by assuming that there is no build up to full production and full production is achieved as soon as construction is complete. This is followed by N years of full production. Hence:

$$C(1+i) = (Rr - FCr - VCr)\sum_{r=1}^{N}1/(1+i)^r$$

This is rearranged to give:

$$Rr = FCr + VCr + K(1+i)C$$

where K is the sum of the geometric series:

$$K = i(1+i)N/[(1+i)N-1]$$

Values of K for various values of i and N are given in Table A6.2:

Table A6.2: Values for K

N	10	15	20	25	30
interest (i)					
5.00%	0.1295	0.0963	0.0802	0.071	0.0651
7.50%	0.1457	0.1133	0.0981	0.0897	0.0847
10.00%	0.1627	0.1315	0.1175	0.1102	0.1061
12.50%	0.1806	0.1508	0.1381	0.1319	0.1288

The capital recovery factor (K_o) is then:

$$Ko = K (1 + i) C$$

and from A6.2, we get:

$$Ko = K (1 + i) Co \sum_{S=0}^{p} a(s)(1+i)^{p-s}$$

Values for the Return on Capital (ROC) or Ko/Co are given in Tables A6.3 and A6.4 for a royalty free basis and one encompassing a 2% royalty to the process licensor, respectively. Table A6.3 has been used for typical non-process items (pipelines, ships etc.) and Table A6.4 for licensed processes.

The selection of a rate of capital return is dependent on many factors including the nature of the industry in question. For upstream oil and gas developments, or relatively small scale process plant, high rates of capital return are often demanded by the investors to offset short operational lives or perceived higher levels of risk. For very long term (30 year) infrastructure projects often accessing government funds, far lower rates of return are required. Many Greenfield operations in the chemicals industry are planned for a lifetime of 15 to 20 years and rates of return are as appropriate. Commonly used values for the return on capital in this work are emboldened in the tables.

Table A6.3: Values for Annual Return on Capital (Royalty Free Basis)

1y construction	N	10	15	20	25	30
	Interest(i)					
	5.00%	13.60%	10.12%	8.43%	7.45%	6.83%
	7.50%	15.66%	12.18%	10.54%	9.64%	9.10%
	10.00%	17.90%	14.46%	12.92%	12.12%	11.67%
	12.50%	20.32%	16.96%	15.54%	14.84%	14.49%
	15.00%	22.91%	19.67%	18.37%	17.79%	17.51%
2y construction	N	10	15	20	25	30
	interest (i)					
	5.00%	13.94%	10.37%	8.64%	7.64%	7.00%
	7.50%	16.25%	12.64%	10.94%	10.01%	9.44%
	10.00%	18.80%	15.19%	13.57%	12.72%	12.25%
	12.50%	21.59%	18.02%	16.51%	15.77%	15.39%
	15.00%	24.63%	21.14%	19.75%	19.12%	18.83%
3y construction	N	10	15	20	25	30
	interest (i)					
	5.00%	14.32%	10.65%	8.87%	7.85%	7.19%
	7.50%	16.92%	13.16%	11.39%	10.42%	9.83%
	10.00%	19.84%	16.02%	14.32%	13.43%	12.93%
	12.50%	23.08%	19.27%	17.65%	16.86%	16.45%
	15.00%	26.68%	22.90%	21.39%	20.71%	20.39%
4y construction	N	10	15	20	25	30
	interest (i)					
	5.00%	14.59%	10.85%	9.04%	7.99%	7.33%
	7.50%	17.39%	13.52%	11.71%	10.71%	10.11%
	10.00%	20.58%	16.62%	14.85%	13.93%	13.41%
	12.50%	24.17%	20.18%	18.48%	17.66%	17.23%
	15.00%	28.20%	24.21%	22.61%	21.90%	21.56%
5y construction	N	10	15	20	25	30
	interest (i)					
	5.00%	14.71%	10.94%	9.11%	8.06%	7.39%
	7.50%	17.61%	13.69%	11.85%	10.84%	10.23%
	10.00%	20.91%	16.89%	15.09%	14.16%	13.63%
	12.50%	24.66%	20.58%	18.85%	18.01%	17.58%
	15.00%	28.87%	24.78%	23.15%	22.42%	22.07%

Table A6.4: Values for Annual Return on Capital with 2% Royalty

1y construction	N	10	15	20	25	30
	interest (i)					
	5.00%	13.87%	10.32%	8.59%	7.60%	6.97%
	7.50%	15.97%	12.42%	10.76%	9.84%	9.28%
	10.00%	18.26%	14.75%	13.18%	12.36%	11.90%
	12.50%	20.73%	17.30%	15.85%	15.14%	14.78%
	15.00%	23.37%	20.06%	18.74%	18.15%	17.86%
2y construction	N	10	15	20	25	30
	interest (i)					
	5.00%	14.22%	10.58%	8.81%	7.79%	7.14%
	7.50%	16.57%	12.89%	11.16%	10.21%	9.63%
	10.00%	19.17%	**15.49%**	13.84%	12.98%	12.50%
	12.50%	22.02%	18.38%	16.84%	16.09%	15.70%
	15.00%	25.13%	21.56%	20.15%	19.51%	19.20%
3y construction	N	10	15	20	25	30
	interest (i)					
	5.00%	14.61%	10.87%	9.05%	8.00%	7.34%
	7.50%	17.26%	13.42%	11.62%	10.63%	10.03%
	10.00%	20.23%	**16.34%**	**14.60%**	13.70%	13.19%
	12.50%	23.54%	19.65%	18.00%	17.20%	16.78%
	15.00%	27.21%	23.36%	21.82%	21.13%	20.80%
4y construction	N	10	15	20	25	30
	interest (i)					
	5.00%	14.88%	11.07%	9.22%	8.15%	7.47%
	7.50%	17.74%	13.79%	11.94%	10.92%	10.31%
	10.00%	20.99%	16.96%	15.15%	14.21%	13.68%
	12.50%	24.66%	20.58%	18.85%	18.01%	17.58%
	15.00%	28.77%	24.69%	23.06%	22.33%	21.99%
5y construction	N	10	15	20	25	30
	interest (i)					
	5.00%	15.00%	11.16%	9.29%	8.22%	7.53%
	7.50%	17.96%	13.96%	12.09%	11.06%	10.44%
	10.00%	21.33%	17.23%	15.39%	14.44%	13.90%
	12.50%	25.15%	20.99%	19.23%	18.37%	17.93%
	15.00%	29.45%	25.28%	23.61%	22.87%	22.51%

Fixed Operating Costs (O)

Working Capital. Rather than capitalise the working capital and handling it with the project capital (Stratton), the working capital is treated as an annual operating cost. The reasoning behind this is that working capital is normally borrowed against the business and is fully recovered at the end of the project. The outgoings are the interest on the debt. The value of working capital can be taken as 5% of the plant capital or 30 days stock. The latter is generally smaller than the former and was used when sufficient data permitted its calculation.

Labour, Maintenance and Administrative Costs. As a general rule, labour and maintenance were each charged at the rate 3% of the capital per annum. For labour, this included both direct and indirect labour costs. For maintenance, this included both materials and labour. Over the past decades, many companies have made attempts to reduce the operating labour and maintenance charges. Labour can be reduced by extensive computer control. However, the success or otherwise, in reducing the maintenance charge is difficult to quantify, several operations have suffered major problems claimed to be due to the cutbacks in maintenance costs. Administrative costs are basically insurance and local land taxes. A value of 1.5% of the fixed capital as an annual charge was used.

Catalysts and Chemicals. Most plants require some chemicals for water treatment purposes. Catalyst charges are based on a 3 to 5 year turnaround.

Other Operating Costs. Some processes require inputs other than the principal hydrocarbon feed. This is usually electric power and typical average values were used.

A7. Indexed Feedstock Costs

For the most part, we are concerned with hydrocarbon feedstock which is related to the prevailing crude oil price. For some feedstock and hydrocarbon by-product, this is a strong linear relationship.

We are also concerned with how construction costs change with time. Work by H. W. Parker[4] has shown relationship in refinery operations between the construction cost index and the refinery fuel cost index. Plotted in the logarithmic form, this relationship has a high linearity with a slope of approximately unity. The relationship is illustrated in Figure A7.1.

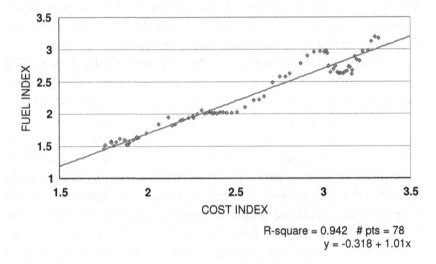

R-square = 0.942   # pts = 78
y = -0.318 + 1.01x

Figure A7.1: Plot of refinery construction cost index against fuel index
(after H.W. Parker)

Since we know that most fuels show a linear correlation with the prevailing crude oil price, we can develop a construction cost crude oil price relationship which is illustrated in Figure A7.2. This shows a correlation plot of the construction cost index against an index based on the price of WTI crude oil. As may be expected, there is more variation in this correlation, but it still shows a correlation factor or nearly 0.9.

Using this correlation we can impute crude oil price corresponding to a particular construction cost index. Using the 2007 value for the construction cost implies an equivalent oil price of $70 per barrel. This is used as the base price for oil and derivatives in the analysis.

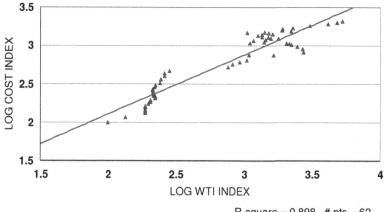

R-square = 0.898   # pts = 62
y = 0.561 + 0.773x

Figure A7.2: Plot of cost index against WTI index

---

[1] Assessment of Cost Benefits of Flexible and Alternative Fuel Use in the US Transportation Sector – Technical Report Three – Methanol Production and Transportation Costs, United States Department of Energy, November 1989.

[2] A. Stratton "A Simplified Method of Calculating product Cost" Technical Note 3, Economic Assessment Service, IEA Coal Research, London 1982

[3] See also T. R. Brown, *Hydrocarbon Processing*, Oct. 2000, p. 93 for a discussion of various cost estimating methods and J. T. Summerfeld , *ibid.*, Jun 2001, p. 103 for an analysis of estimation accuracy.

[4] H.W. Parker, *Oil & Gas Journal*, Aug. 4, 2008

INDEX